MW01248546

Different Waves,

Different Depths

Stories

Different Waves,

Different Depths

Stories

Different Waves,

Roy Christopher

Different Depths

DIFFERENT WAVES, DIFFERENT DEPTHS

First edition

Set in Verdigris Text and Atlas Grotesk
Cover and text design by Patrick Barber
Design for "Fender the Fall" based on the original
novella design by Mike Corrao

ISBN-13: 9798988013303
Library of Congress Control Number: 2023938128

Impeller Press books are published by Patrick Barber in Portland, Oregon, which was built on top of village sites of the Multnomah, Wasco, Cowlitz, Kathlamet, Clackamas, Chinook, Tualatin, Kalapuya, Molalla, and many other tribes who made their homes here. Today, members of these tribes and others are part of our community in the Portland metro area. We honor them with this land acknowledgement, and we give 5% of our gross sales to support Indigenous communities and the Landback movement. Please visit landback.org for more information on the Landback movement.

This book is printed to order, just for you. Thanks to this remarkable manufacturing technology, tiny publishers like myself are able to make books widely available without the cost or waste of printing, shipping, and warehousing books we don't need.

impellerpress.com

Printed and distributed by Ingram Publishing Services

Dedicated to the memory of Kelly Lum

Table of Contents

Drawn & Courted

"THERE'S ONE MORE. YOU GRAB IT, AND I'LL pour us some wine," says Katie, a sturdy brunette with a bob and a widow's peak.

"Okay," Elle says sheepishly through a wall of strawberry blonde hair.

Katie grabs her arm. "Hey, it's going to be okay. You're going to be okay. Mi casa es tu safe place." Boxes are strewn around the living room of the third-floor, two-bedroom walk-up. Katie picks up the remote and turns on the TV as she heads to the kitchen.

Elle swings her hair aside and tries to smile as she walks back out the door. She grabs the last box out of the back of her car and closes the hatchback. The sun is setting on the otherwise quiet street.

"*You saw them on the bus, but you didn't get to talk to them. You've taken the same bus at the same time for weeks, but you've never seen them again.*" The voiceover is unignorable. Animated infographics dance on the screen. "*Or . . . You've had The One in mind for years, or you woke up one day with the perfect vision of a mate, but you've never seen them in real life.*"

Elle trudges back in and drops the last box. She sits on the couch and puts her cellphone on its arm. Just then Katie comes back in with an open bottle of wine and two coffee mugs.

"*Missed encounter or dream soulmate, your vision becomes a reality on* Drawn & Courted."

A logo with a pencil and a rose crossed over a heart emerges on the screen. Katie and Elle look at each other wide-eyed then quickly back to the TV. Her eyes locked onto the screen, Katie pours wine into the mugs. Elle picks up one of them and settles into the couch.

"*It's happened to every one of us: We see our dream mate on the bus, in a store, or on the street. It's love at first sight, and then the vision ends. Or... We see them in our dreams: We have the perfect mate in mind, but we've never seen them* IRL.

"*Either way, our team can make the dream a reality. Here's your host, Dean Crispino!*"

The studio audience erupts in applause as a man in a crisp navy-blue suit jogs out on stage. He has a swoopy-banged, Tony-Hawk haircut and a microphone headset on.

"Good evening and welcome to *Drawn & Courted*, where we connect the ones who missed and find the ones hiding in dreams. I am Dean Crispino. Let's meet our first contestant!"

The camera swings to a nervous-looking man in his late 20s. His ill-fitting jeans pool around a new pair of New Balance sneakers, and his crumpled button-up oxford looks as uncomfortable as he does.

"*Marcus Ozechowski is a marketing manager from Tacoma, Washington. He's a basketball fan who likes music from the 1990s.*"

"Holy crap! What's he doing on here?" Katie says to the TV.

"I don't know," Elle says.

Dean waves Marcus to center stage.

"C'mon over here, Marcus!"

Marcus walks on stage, hesitantly at first. Dean checks his teleprompter.

"A basketball buff who likes 1990s music... So, I guess Pearl Jam is your jam?"

"Well, not exactly. I'm more into—"

"No time, Marcus. Let's get to making that connection you missed. Here's how the show works!"

The camera swings again to a screen showing animated infographics. They fill the screen as the voice-over returns.

"*First, we take your vision—real missed encounter or imagined soulmate—and we render it real. Our professional sketch artists will bring your mate to life by drawing them to perfection based on your detailed description.*

"*Second, we run their drawing through our state-of-the-art facial-recognition database, locating your dream mate in the flesh.*

"*Third, we contact them and send the two of you on the first date of your dreams!*"

The animation stops on a bulleted list as the voice-over recites them in turn:

"*Vision. Drawing. Recognition. Reality. It's* Drawn & Courted*!*"

The audience applauds as the camera cranes back to Marcus, now seated across from a woman with a tablet and a stylus. Dean stands behind them on the stage.

“Please welcome one of our talented sketch artists, Laura Ancona!”

Laura waves. The crowd claps.

“*Laura is an illustration graduate of San Diego State University and spent the five years since as a forensic artist at the Los Angeles Police Department where her sketches have led to hundreds of arrests!*”

More applause from the in-studio audience. Dean nods and smiles as if he didn’t know this about Laura.

“Impressive! Now, Marcus, you saw someone you’d like to see again, correct?”

“Yes,” Marcus sits up straight. “I was in the Comet up on Pike in Seattle, and I saw her at the bar. It took me most of the night to get up the nerve just to talk to her, and she was leaving when I finally did. I didn’t even get her name.”

“Aw . . . Sad story, Marcus. Now, describe the woman from the bar to Laura in as much detail as you can remember.”

“Well, she had strawberry blonde hair and big blue eyes . . . She was a full head shorter than I am . . . ”

Laura scribbles on the tablet. Dean is suddenly impatient.

“And how tall is that, Marcus?”

“Oh, sorry. I’m six-two.”

“Her face, Marcus. What did her face look like?”

“She had the big blue eyes, like I said, and the reddish hair . . . ”

"The eyes and hair are good, but Laura needs to know more specifics about her face: the shape, skin color, eyebrows, freckles, moles . . . Any specific details you can remember about her face."

"She looked kinda like Meg Ryan but with the reddish hair."

"My god. I met him at the Comet. Is he talking about me?" Elle asks out loud.

"If so, that was generous," Katie says.

Elle doesn't react. She stays focused on the screen.

Dean turns to Laura, who is still scribbling on her tablet.

"Is that quite enough to move us along, Laura?"

She makes a last few flourishes and turns the tablet toward Marcus.

"That's her!" he says.

The drawing fills the screen.

"That's me!" Elle says at the TV.

"Holy shit!" Katie says at the TV.

A relieved Dean says, "We'll bring in the biometrics—Right after this!" He points at the camera as it cuts to a commercial break.

Katie mutes the TV and turns to Elle. "What is he doing?"

Stunned, Elle takes a big gulp of wine from her mug. "I don't know."

"Didn't you stop returning his calls literally right before you decided to move in here with me?"

"Yeah," Elle says blankly. Elle gets up, dazed, and walks toward her new room. "I need to unpack."

"Don't you want to see what happens?" Katie asks as she turns toward the TV.

"No, it's too weird!" Elle says from her room, "And kinda scary." She closes the door.

"They're on TV! In a studio! In L.A. somewhere!" Katie yells.

"You don't think it's only a matter of time before producers or whoever come knocking?"

"They can't know where you are!"

"If you say so!"

Katie turns back toward the TV, unmuting it as she leans its way. "Let's see if it works."

"We're back on *Drawn & Courted*. I'm Dean Crispino, here with our forensic artist, Laura Ancona, and one half of our missed connection, Marcus Ozechowski—"

"I go by 'Ozzy'—"

"No time, Marcus. Now, how does this work?"

"*The National Institute of Standards and Technology maintains a list of the best facial-recognition algorithms in use,*" the voiceover continues on cue. The screen shows the sketch on one side and a flickering procession of faces on the other with innumerable illuminated green match points flashing on each picture. "*We use a combination of* NIST*-approved software and a* *process of data conflation to get the most accurate results available anywhere. We currently use the five largest databases of faces, over 50,000 points of contact, and operate at a 99.8% accurate recognition rate."*

Just as the explanation ends, the picture flashes a complete match: Elle's face covered in tiny green dots.

"OMG," Katie mutters to herself.

The green dots disappear as Elle's face and full name fill the screen.

"Well, we did get there," Dean says to no one during the next commercial break. He tears off his headset. Rubbing his eyes and pushing his bangs back dramatically, he yells, "Can I get a shot of something before we're back?"

Marcus is shuttled further backstage by a production assistant as another hands Dean a cup of water. He sets it on the counter backstage. Another PA comes by with a glass tumbler and a bottle of bourbon. Dean grabs it before the PA can set it down, pours himself two fingers, and throws it back. Dean's eyes roll back with his head. He shakes it off.

"Much better," he says. "What's next?"

Katie turns her phone over on the arm of the couch, just in case. "Come watch the second half," she yells at Elle's closed door. "It might take your mind off the first."

"No, thanks."

"The next part's supposed to be different. It'll be fun!"

"Enjoy!" Elle says.

"You know, I went out with Ozzy before you did."

Elle doesn't answer.

"We never got past the first—it's back on!"

"*Up next we have Frank Bean. A logistics engineer for a major online retailer, Frank has a passion for prompt deliveries and hardcore hip-hop.*"

Dean bounds back onto stage where Frank is sitting across from Laura and her tablet.

"Welcome, Frank!"

"Thanks, Dean."

"Are you ready to meet the girl of your dreams?"

"I am, Dean."

Katie's eyes grow wider and wider as, detail by detail, Frank describes her perfectly. "Oh my god!" she yells.

"She has short, dark hair, a widow's peak—"

"What is going on today?" Katie says to the TV.

"—almond-shaped, brown eyes."

Katie is rapt to the screen as a sketch of her face emerges. "This *is* too weird," she stammers as she turns off the TV.

Dean throws back another shot of bourbon as Frank is escorted backstage. A software engineer approaches, gingerly holding a tablet to his chest.

"I think we have a problem, sir."

"We have all the human behavioral data in the world at our fingertips, from sexual orientations to sexual fantasies, from toothpaste preferences to application preferences, from family history to browser history, from genetic makeup to, well, makeup...And yes, of course, facial features. How can there possibly be a problem?"

"The results of the last two rounds are apparently acquaintances. Not only that, but they live in the same place in Seattle."

"Not to worry, we've had results from the same city on the show before," Dean pours himself another two fingers and raises the glass.

"No, sir. They live in the same apartment," the engineer adds, turning his tablet toward Dean.

Dean stops before the glass reaches his lips. "Wait, they're sitting on the same couch, watching the show together right now? Is that what I'm seeing? Why didn't we know this *before* show time?" Dean pleads, putting the bourbon back down. "More importantly, if you're down here, then they know about this upstairs, huh?"

The engineer nods. They stand in silence for a moment. "Awkward day for data, huh, sir?"

A little while later, a man of nearly indeterminable gender or age, wearing all black and carrying a tablet, addresses a conference room full of engineers and producers, including Dean and the engineer.

"How exactly did we arrive at these results? What were the datasets?" the man says.

"For Elle, Marcus told us a lot," another engineer says, "but we got her information just using her phone number to tap everything the system had already collected from her phone. For Katie, we used Frank's browser data and pornography preferences cross-referenced with the demographic and psychographic data we gathered from women in and around his zip code and former delivery routes."

"Don't we have fail-safe measures in place for just these kinds of contingencies?"

"Yes, sir, but they only work if the data is kept current. Apparently, neither woman had updated her address—anywhere—and their lease hadn't hit our database yet."

"Our whole enterprise is based on disrupting the normal flow of information and collapsing contexts either pushed apart by consequences or kept apart by consciousness," the man says. "I know we're basically just illuminating latent networks, but these sorts of social and geographical overlaps are supposed to be inherently hard-coded out, systemically avoided. So, what happened out there today?"

An engineer raises her hand, and he nods at her. "Our problem appears to be the world's problem, sir. It's a global issue."

"How's that?"

"The networks are saturated. Everyone on earth is connected now by at least a Granovetterian weak tie."

The man crosses his arms. Dean throws his hands in the air. "What does that mean for us, for the show—bearing in mind that I only need an actionable level of understanding."

"It means that we're going to have a more and more difficult time connecting two people who don't already know each other somehow," she says. "All of the nodes are connected. All of the contexts have already collapsed."

"More importantly—" the man starts as another engineer opens the door, knocking as she steps in.

"Yes?" the man says to the engineer at the door.

"We have a problem, sir."

"Another one?"

She looks at her tablet. "Yes, sir. We've been hacked, and we have a leak."

"Great! How big of a leak?"

"A big one. A WikiLeak."

Dean leans over to the producer next to him. "What was that other game show idea you had?" he whispers.

"'Accusers vs. Abusers'? It's a *Family Feud*-style show, where—"

Dean stops her. "I think I get it."

Katie finally turns off the TV for good. Then the house phone rings. Elle comes out of her room as it rings again. As it rings the third time, Katie's cellphone buzzes.

On the tablet of one of the engineers in the conference room, from flickering black to a blurry POV pan of carpet and a couch, Katie's face finally fills the screen.

Katie is looking at her phone. The screen reads "Unknown Caller."

Then the doorbell from downstairs buzzes. Elle reaches to turn off her phone, and it immediately rings. Then there's a knock at the door.

Outside the apartment building, a line is forming. Men with flowers, adjusting their attire, checking their breath. More emerge from cars, as others pull in, looking for parking.

Kiss Destroyer

WE MET HALFWAY. FOR THE FIRST TIME SINCE meeting her, I knew definitively that she was with someone. She was engaged. The wedding was a few months off. We talked and we drank and we danced and it felt like it always felt. I was overwhelmed. The only thing that kept me grounded was knowing that in a few months, she'd be married to someone else. And I'd be gone.

I leaned in close to her ear and whispered, "This is nice."

She stopped, stunned. She flashed a withering look and edged away from me through the crowd.

"Wait!" Hearing me behind her, she hurried on. I caught her in the bar. "I meant that it felt nice knowing—"

"No, I feel the opposite," she turned and said. "It doesn't feel nice knowing. It feels awful!"

"Well, I was speaking for you. I thought—" She put her finger on my lips to shush me. She was definitely angry but seemed ready to recover.

"Want some?" she asked, pulling a flask from her purse.

"What is it?"

"Have some or don't," she said over her shoulder, walking out onto the balcony.

"I didn't think—" I said as she drank.

"You always knew." She handed me the flask. I downed a gulp of sweet liquid. It tasted the way antifreeze smells, perhaps a flavored vodka of some kind. "I always hoped, but I never knew."

"Is that why you're here now, hope?"

"Yes."

"Well, all of your hopes are here, and they're all shit. Sorry."

As I took another swig, everything took on a fog, soft around the edges. I felt anger and disappointment sharpening in me. "Then why are we here? What is this?"

"Let's dance!" She said, draining the flask.

"I don't want to—" She grabbed my arm and dragged me inside. She kissed me deep, hard, obviously feeling the drink, and then pulled me onto the dance floor.

The music and the bodies blurred. We were together, then apart, then together. One minute, we were blended into one, the next, we were on different planets. Other bodies remained distinct, but ours melded and folded and separated like taffy. The music was one long song, and it was always exactly the right one.

The melding continued when we finally made it upstairs to bed. I'm not even sure we had sex, but we were one many times over before we slept. We fell in and out of love over and over, fighting, folding, fucking. I wish I could remember it more clearly.

"Every time you make a decision, it's like destroying a whole other world," she told me earlier that evening. "I never wanted to destroy this one."

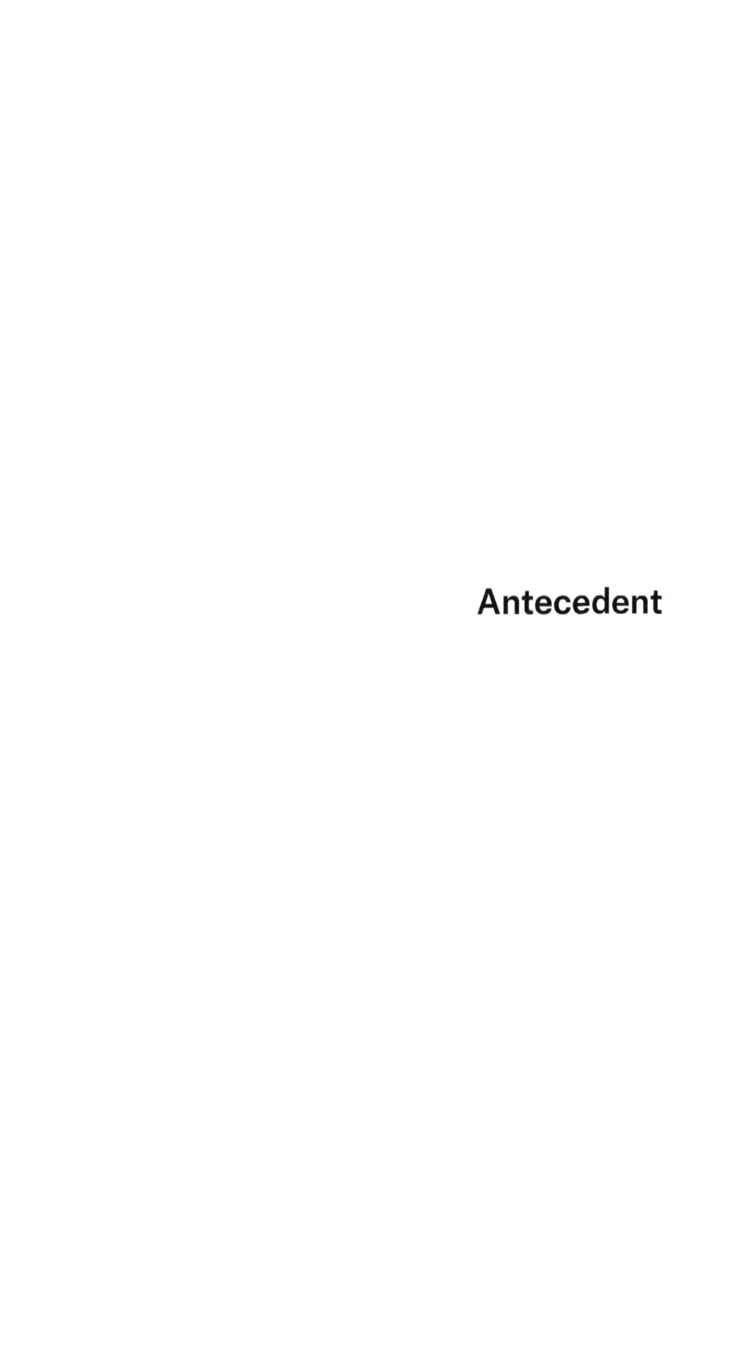

Antecedent

HE HELD SIX DISCRETE BITS OF INFORMATION in his head simultaneously. No way to write them down so he recalled them in order on three separate occasions, six blanks in formation like a constellation. He awoke with only the memory of remembering, but not the information, his head a form unfulfilled.

An hour later, he was up and reading a well-worn library copy of Debord's *Panegyric* over coffee. He put down the book and ran his hands through his greasy hair. His stomach grumbled. "Time to venture out," he thought as he grabbed his jacket. His one-room apartment was strewn with similar texts: *The Revolution of Everyday Life*, *Temporary Autonomous Zone*, *The Society of the Spectacle*, *Exploits & Opinions of Dr. Faustroll, Pataphysician*. Not much to do but read when you're chronically unemployed and can't get away with stealing anything, let alone a TV or cable. He'd done several short stretches of time for petty theft, check fraud, and impersonating a police officer, but it had been a while since the last job and the last bid. He fancied himself a revolutionary, a civil disobedient. He downed the last half of a cup of coffee, rinsed it, refilled it with tap water, and downed that. Then he grabbed his jacket and headed for the stairs.

Walking by a gallery down the street, he watched as an older gentleman loaded paintings into the space from a beat-up blue Econoline.

"Can I help you with those?"

The man barely looked at him, but nodded.

He grabbed the next one from the van and walked it longways through the narrow doorway. The man he was helping never said a word as they filed the covered canvases from the van into the small gallery.

Walking in with the last one, he heard the van's engine stutter on. Figuring that the man was moving it to a proper parking spot, he delicately ripped off enough of the packing paper from one of the canvases to see the work. He was aghast at the sight of it. It was like nothing he'd seen before. The colors, the shadows, the rhythms sang to him like a bird uncaged. He was seduced completely. If he were able to paint, this would be the statement he would make. Unable to help himself, he ripped open another one. The effect was repeated twofold. He began to sob as he set the second one upright. This one was larger and nearly vacant, having been washed over several times with white gesso. The few extant strokes and colors were exactly as he would have wanted them. He tore open another.

Lost as he was in the collection of art he had arranged along the walls of the space, he failed to notice that the man he'd helped hadn't returned. The paintings' spell remained unbroken by the voice of the curator approaching.

"Hello... Sir?"

He hadn't heard her, still examining the six paintings leaning against the walls, surrounding him: a tryptich along the longest wall, two tall ones opposite, and the biggest one on the back wall by itself—the blanks in order, just as he had remembered.

"Thank you so much for letting us show your work. Do you have an arrangement in mind? We'd like to get them up as soon as possible. The reception is in a couple of hours."

Still trying to compose himself, he didn't speak at first. Then, pointing around the three walls at the paintings there, he said, "Just as they are now."

"Also, are you sure you don't have anything in mind we could put on the marquee? We've left your name off as requested."

"Thank you. That'll do."

"Okay... Will you be joining us tonight?"

He finally turned toward her. "I'm really not sure yet."

She forced a smile. "Well, I hope to see you in a bit."

He was hesitant to leave the paintings behind but stepped out into the evening air with a buzzing heart and head.

He stopped by a newsstand to check the paper for an article about the show he'd just taken responsibility for. Looking back through the weekend edition, he found it.

"Renowned Artist's First Show in Over a Decade," read the headline. The article said that the paintings had been created during nearly twelve years of total seclusion. After years of acclaim during which the artist felt he could do no wrong, he grew disillusioned with the art world. He decided to isolate himself and dig deeper, attempting to create a completely new form of painting.

"I'd say he succeeded," he said to himself as he closed the paper. His stomach groaned again as he wondered if there would be free food at the opening.

Back at the gallery, the paintings were mounted just as he'd asked. The marquee out front read simply, "Artist's First Show in Twelve Years."

He walked in with borrowed confidence, but his smirk disappeared as soon as he saw the paintings on the walls. It was difficult to contain his emotions in the presence of such passion, such control. He stared into the strokes, mesmerized.

The curator walked in, startling him. He quickly stacked some cheese on a cracker and poured himself a glass of wine. Following his lead, she did the same.

"So good of you to join us," she said over her glass.

He turned, finally acknowledging her presence.

"There was some doubt that any of us would get to actually meet you, given your lengthy seclusion and request to be alone while bringing in your work."

"Well, I changed my mind," he said smiling.

"The show is everything we expected and more," she said looking at the two paintings on the wall nearest her, but when she turned back around, he wasn't there. There was only his half-empty wine glass and half-eaten cheese and cracker tumbling to the floor beside her. “Where’d you go?”

“I’m right here,” he said, one painting to another, but she couldn’t hear him. He didn’t recognize his own voice. The peace in the paintings became panic in his veins.

Six discrete bits of information hanging in his head simultaneously. No way to write them down so he became them, six paintings in formation like a constellation. He awoke with only the memory of remembering, but not the information, his head a form with empty blanks. Six blanks now staring out from three walls, their form fulfilled.

Not a Day Goes By

I ADJUSTED FOR THE WIND AND EVERYTHING. The kite veered way too far off the southwest corner of the building anyway. The more I tried to get it out of the flight path, the more adamant it seemed on staying there, a Jolly Roger taunting me from the no-fly zone.

Its flapping black plastic was overtaken by blades chopping air. The sky is quiet at seventeen stories up, and you can hear a helicopter coming from a long way off. No matter; my pirate prey remained in harm's way.

I cut the line at the last possible second and watched as black blades and black plastic met in a violent twisting tryst. Running for the stairs as the helicopter sat down, I heard yelling as I hit the second landing. With my hood up, I knew they wouldn't identify me later. I stopped a few floors down, shoved the hoodie in my backpack, put on a hat, and headed for the elevators.

I woke up in the night just enough to see her lying across from me. We had collided into each other avoiding a messenger while crossing at Denny Way. I was running down Olive. I'd just stolen coffee from Coffee Messiah as I do every evening. Mr. Turner and most of his employees hate me, but they'll never know who I am. On any other night, I would've ducked into the alley behind Dino's to drink my coffee and figure out what to do with the rest of my night. That night I decided to cross the street.

It was November 15, 1997. I know because the clock resets at 11:59 every night, as it has every day since my 23rd birthday. As far as I know, everyone else moves on through the calendar. When I wake up, it's November 15, 1997 again, and I am wherever I was at midnight the night before. In the 1,207 November 15ths since 1997, I've met other prisoners of the day. One was so obsessed with fixing the problem, she wasted the day over and over again. Another was so bent on revenge that she spent every day getting back at everyone who'd wronged her up to that day. Another just couldn't take it and killed herself over and over. I couldn't be around any of them for very long. I was pretty sure this girl was one of us, but I needed more time.

When I crossed Denny, I didn't see her until she almost knocked me down and nearly fell herself. I caught her and we swung out of the path of the speeding bicycle. I spilled my coffee all over both of us.

On November 15, 1997, the Leonid meteor storm was gathering force. The moon was nearly full, waning from its full phase the night before. Bill Clinton gave his weekly presidential radio address. Crime was way down. It was the Day of the Imprisoned Writer. It was the 319th day of 1997. It was also a Saturday and the first America Recycles Day.

I don't know what information is relevant, but I know the day well. Call it a time loop, a flat circle, the eternal return, recurrence, or repetition, or just Groundhog Day, it had been all the same for 24 hours over 1,200 times. Then she showed up.

I mentally retraced my steps. Had I done something wildly different today? See, the irony of the loop is that though I'm stuck in the same day, the day itself is the same for everyone except me. I can cause changes, but they all reset by midnight. Most things stay the same. All of that to say that I should've known she was coming across the street unless I did something earlier in the day to cause her to change her course this time. Where had I been all day?

I got up that morning and got my usual First Church breakfast. I took a 43 bus downtown and walked to the Vashon Island ferry. I wanted to go to the bookstore and have lunch this burger place on Vashon Highway. I got caught trying to skip the ferry toll and had to sneak on with the cars. Maybe she was in one of them.

I lifted a new hardback copy of *Great Apes* by Will Self from the bookstore and walked to Perry's Vashon Burgers. There I ordered a garden burger, a small order of fries, and a vanilla shake. I started the book, which is about a man who wakes up in a world of apes and thinks he's the only human, but I was distracted by the marquee on the Vashon Theatre across the street. *Midnight in the Garden of Good and Evil* was playing. One of those brief waves of déjà vu hit me, like several coincidences piling up together in the same moment. I shook it off as my food arrived.

When I got back to downtown Seattle, I walked most of the way back up the Hill and made my coffee run. I was crossing the street to go to Twice Sold Tales. That's when I ran into her.

"I'm so sorry," I said as we twirled back onto the sidewalk.

"Where are you going in such a hurry?" she asked.

"I was going to trade this in." I pulled my copy of Tibor Fischer's *The Collector Collector* out of my backpack.

"Oh, that's a good one! Why are you trading it in?"

"Good question. I thought about this earlier. When I trade this in, I'll likely get $3. That will cover this evening's coffee."

"Right," she seemed more interested than I'd expected.

"But if I saw this book for sale, for twice that, which is what they'll put it out for, I'd probably buy it."

"Me too."

"So, what does that say about my relationship with this book? Shouldn't I just keep it?"

"Maybe, but it's like you don't want to have it, you really just want to buy it again," she said, satisfied.

"Insightful," I agreed, nodding, "but what does that say about me? That's the part I've been trying to figure out."

"It seems like you're stuck. You're hung up on the beginning, that feeling of newness."

"Wow..."

"I can relate," she responded quickly, as if trying to hinder an uncomfortable pause. "You can't get it back though. It's the entropy of experience. I want to buy the first Bad Flag record every time I see it. I know it will never give me the same feeling again, but I can't help myself." She looked at the book again. "Mind if I tag along?"

"So, what stage are you in?" she asked, petting a grey tabby at Twice Sold Tales.

I didn't look up from the book in my hand. "What do you mean?"

"I know your situation, and I know the stages, so which one are you in?" She insisted. I still didn't answer. "There are five emotional stages of dealing with time loops. Which one are you in?"

"What?" I finally acknowledged, looking at her.

"You're in a funk. Which Red Hot Chili Peppers guitarist are you?"

"I still don't know what you're talking about, but I'm Frusciante, for sure."

"Acceptance," she said, as if analyzing me and jotting down my answers.

"What are the others? What is Hillel Slovak, for example?"

"Slovak is Depression. Dave Navarro is Anger, obviously."

"And the others?"

"Well, Jack Sherman, Arik Marshall, DeWayne McKnight, and Jesse Tobias are collectively Denial."

"If they're all taken as one, that leaves one more. Who's the other? Who is Bargaining?"

"Josh Klinghoffer," she said finally.

"I don't know that one," I said.

"You wouldn't."

"Whoa there, hipster lady! I know more about music I hate than you do about music you like."

"Is that right?" She put her hands on her hips.

"Yep."

"Well, let's just say I'm ahead of you on this one." Her expression suddenly turned serious as we got ready to leave. "I saw the kite."

"What kite?" My immediate reaction.

"Well, I saw the story first. That's how I know your situation."

"What story?"

"The *Post-Intelligencer* story about the guy supposedly stuck in the same day."

"'Supposedly'? I guess they didn't believe me. I never got to see the story, for obvious reasons."

"Here, I brought it back with me." She opened her backpack and pulled out a crumpled print-out of the Sunday edition of the *Post-Intelligencer* from November 16, 1997. "At first we thought it was another false fire, but then we saw the kite."

"'His case has baffled and intrigued doctors who examined the 23-year-old'," I read aloud, "'who first experienced the sensation shortly after he started at the University of Washington, because he does not exhibit any of the other neurological conditions usually associated with those who suffer from déjà vu. UW Psychology professor Dr. Christina Kopinski thinks that anxiety is causing the appearance of repetition in his brain—anxiety that may have been exacerbated by the man dropping out of school. "The general theory is that there's a misfiring of neurons in the temporal lobes, which deal with recollection and familiarity. That misfiring during the process of recollection means we interpret a moment in time as something that has already been experienced," says Kopinski.' . . . *For over three straight years?*"

"They didn't believe you, but I do. It's called *déjà vécu*, 'already lived through.'"

"I don't care what you call it! I want out!"

"Shhhh!" the guy behind the counter urged as we reached the door.

"A little too Navarro there," she added.

"We don't completely understand it yet," she started as we walked outside, "but it usually has to do with trauma. It's a never-ending meal. It ends up on your plate, and you have to eat it, over and over, every day. It's a loop that won't close. It doesn't feel like it has happened. It feels like it's still happening."

"So, which is it? I want the same feeling again or I don't?"

"You're an extreme case. They seem to be the same thing with you."

"How is that?"

"You're both abortively resigned to your day and pregnant with retaliation for it," she said solemnly. "We've never seen such extreme poles in one case."

"How do I get out?"

"That's why I'm here."

When I finally woke up for good the next morning, she was gone, but it really was the next morning. November 16, 1997. I know I'll never get that feeling again, but not a day goes by that I don't wake up and wish she were still here.

I checked my watch. There was probably time to spill more coffee on my pants. Maybe even time to drink some.

Dutch

[NOTE: THE FOLLOWING DOCUMENTS WERE included in the liner notes to Bad Flag's history-spanning boxset, *Dutch*, released by Numero Group in 2019.]

FIERY TALE: A BRIEF HISTORY OF BAD FLAG

A work of art has to exist in a world as an object, as real as the sun, grass, a rock, water, and so on. It must also possess a slight error. In other words, to be right, it has to be a little bit wrong, a tad strange, and thereby, truly real.

—KHARMS—

"They were the best," says one fan with the measured reverence usually reserved for religious worship. "It's really too bad more people don't know about them."

"No, it's not," another counters. "I'm glad no one knows about them. People ruin things."

The latter fan expresses the prevailing attitude of the underground's old guard. It's a mixture of ownership, selfishness, and elitism that says, *This is ours, and you can't have it. You don't deserve it.*

Whether or not you agree, there is a certain cachet that is diminished when something gets too big. When everyone knows about a cultural phenomenon, its allure is lost. Cool doesn't scale.

Though they were once approached by the A&R of a major label, Bad Flag were never in danger of getting too big.

"I saw them open for someone at the Off Ramp in Seattle in 1994," Nils Bernstein tells me. "I can't remember who it was. They were on their way out by then I've been told, but you certainly couldn't tell by the way they played. They erased the headliner from my head!" Bernstein worked in publicity at Sub Pop Records from 1991 to 1997. At the time, Sub Pop was expanding rapidly. Flush with Nirvana revenue, they were busy diversifying their roster and had recently signed a wave of successful indie bands, many at the behest of Bernstein. "On their way out or not, they're still easily the best live band I've ever seen—and I've seen a lot of bands ... The records really don't capture the lightning of their live show." He trails off. "Hands down. The best."

Bad Flag broke up in December of 1994, so Bernstein saw one of their last shows. The records he's referring to, of which he has two, are a scant series of seven-inch singles the band put out. Depending on who you ask, the sum of the band's output is either three or five seven-inches, one demo cassette, a rehearsal tape, and two live bootlegs. Some say the demo is just a bad copy of their early singles, but no one can explain the extra song that only appears there.

"I picked up the only two records they had with them at that show," Bernstein continues. He pulls the two records out of a steel box to the side of his racks of vinyl. I try to see what else is in there, but he quickly closes it. He holds out the two pristine artifacts, both

sheathed in thick plastic sleeves. "I played them both only once, and only then to record them to tape. I mean, okay, I had to play 'All the Way Down' twice because I forgot to un-pause the recording, but that's it."

"All the Way Down" is the third act, the B-side to the three-song seven inch, *Man Amok* (Touch and Go, 1995), which marked a conceptual turn in the band's songwriting. Acts 1 and 2, "Where the Day Goes to Die" and "Good God Gone Bad," are on the A-side. It's probably their best-known trilogy of songs. It was their last, and the only recording not released by the band themselves—or without their knowledge. The change is subtle but significant enough to make one wonder what would have come next.

"They didn't really seem interested in selling them," adds Bernstein, "and when I told them who I was, they packed up and retreated backstage. I didn't even get a chance to bring them to the attention of the label."

WAVING RADIANT: WHO IS BAD FLAG?

If ever there were a band deserving of the designation *power trio*, it is Bad Flag. Their music is a mathematics, an algorithm. It's a process in progress that they are neither enjoying nor enduring but exacting, like an angry surgeon. It's as heady as it is heavy. The three in question are the affable oaf, Dutch McNeal (drums), the cryptic yet quotable Sam Sports (bass, vocals), and the even-keeled Will Wilson, Jr. (guitar, backing vocals).

Don't let their name and logo (the skull and crossbones of the Jolly Roger) fool you, their lyrics are written and sung by a young man worldly beyond his home and wise beyond his home-schooling. In one song, he can go from lines about "the stress-free skin of the unperturbed" to a string of expletives lacking rhyme, rhythm, and reason. One reviewer wrote of their first 7", "No one should have it, and no one should be without it. That's how controversial it is."

Bad Flag was formed in the forges of Chicago during a particularly hot time. "Sam and I have known each other since middle school," Will tells me over the phone. "We met Dutch in college." By *college*, he means the University of Illinois-Chicago. After forming, dissolving, and playing in several other bands throughout school in the Chicagoland suburbs, Sam and Will moved to Wicker Park together, hell-bent on starting something new. Paul Morley's summation of Joy Division from June of 1980 could just as easily have been written about Bad Flag:

> *Good rock music—the palatable, topical stuff—is an amusement and an entertainment. But the very best rock music is created by individuals and musicians obsessive and eloquent enough to inspect and judge destinies and systems with artistic totality and sometimes tragic necessity; music with laws of its own, a drama of its own. The face of rock music is changed by those who introduce to the language new tones, new tunes, and new visions.*

"We had a vision that never manifested in any of the other bands we'd been in," says Sam.

"It wasn't like we were trying to start a revolution," Will adds, "but we were trying to realize something we hadn't heard anywhere else, for ourselves." Their vision included starting with the basics—vocals, guitar, bass, and drums—and building up. "We wanted the constraints of a regular band," says Will, "but we wanted to push on them as hard as we could."

And push they did. Dutch seemed the missing piece. His drumming is propulsive, steady yet organic. He set up his kit with his back to the audience, and while some found this impersonal, Dutch was actually trying to be as close to the action as possible. As precise as he could be, Dutch was not a machine. He was an animal.

Sam's songwriting might be the thing everyone remembers or writes about, but Will and Dutch are essential. "Bad Flag is the three of us," Sam insists. "And no one else."

Bad Flag emerged on the Chicago club scene seemingly fully formed. Their first show, opening on a three-band bill that included their heroes the Jesus Lizard and local powerhouse Tar at the Lounge Ax in 1991, was plagued with sound problems, but they played almost flawlessly. It was a performance that didn't go unnoticed.

"We weren't ready for them," says Tar's John Mohr. "I thought *we* were bound by tension! Those guys seem ready to snap the second they plug in."

"There was a lot of amazing music in our circles at the time," Steve Albini remembers. "Tortoise, Tar, Naked Raygun, The Jesus Lizard... Brise-Glace, anything with David Grubbs in it, or Jim O'Rourke... It was hard to stand out, but Bad Flag was a revelation." Albini ended up recording all of their official releases. "I spent every session trying to recapture the magic of that first show," he claims.

"I think he came pretty close a couple of times," says Corey Rusk, owner of Touch and Go, who put out the band's last 7" record. "Even on the later material, which we were over the moon to release, where they stretched out more than ever before... What a band..."

Though those few recordings are bought and sold like gold, everyone knows these rare documents don't capture the caged beast that was Bad Flag live. In between that first Lounge Ax show in 1991 and the posthumous *Man Amok* 7" in 1995, there were two brief tours in the Anything Grows flower shop van. In 1993, Bad Flag headed east, and in 1994, they headed west.

"The tour in 1993 was mainly to go to DC," Will tells me. "We felt like, outside of Chicago, Olympia and DC were where our kindred bands lived. So, we booked that tour aiming to spend a couple of days in DC. We'd sent Ian [MacKaye of Fugazi and Dischord Records] our records, and he got us on a show at the 9:30 Club with Jawbox." They also played with Five-Eight at the 40 Watt in Athens, Mary's Pet Rock at The Nick in Birmingham, Polvo at Cat's Cradle in Carrboro, and scattered shows by themselves in between.

Their second tour was in the other direction. Setting out for the Cascades, they seem to have found a second family in the Northwest.

"We bonded instantly with a lot of the bands out there," Will says. "I mean, we have at least as much in common with Hush Harbor and .30-06 as we do with Slint and Gastr del Sol." It's true. As much as Bad Flag fit in with the Chicago bands they played with, it's not difficult to imagine them coming up in Portland, Olympia, or Seattle.

"Some Velvet Sidewalk? Unwound? Lync?" Dutch added. "We could've easily been on K Records or Kill Rock Stars and no one would bat an eye."

The band ended with that tour. They announced their break-up before they got back. The final performance at the Fireside Bowl in Chicago was an emotional affair. Thankfully, someone recorded it. Like seemingly every show, they played it like it was their last. This time they were right.

BAD FLAG INTERVIEW, MARCH 17, 1993:

The following brief interview was conducted shortly after Bad Flag released their third seven-inch record, "Get Off the X" b/w "Dogspine" (Vortex Shedding, 1993), and just before they went on their first tour.

So, are those your real names?

DUTCH: Yep.

WILL: They're real enough.

"Sam Sports," really? You don't seem like the sporting type.

SAM: 'Irony' is my middle name.

DUTCH: It is!

WILL: He legally changed it.

Sports seems like more of a Dutch McNeal thing.

WILL: He played in high school.

DUTCH: Yep. Linebacker.

What about the band's name, "Bad Flag"?

WILL: That was also Sam's doing.

DUTCH: Yeah, blame Sam. [laughs]

SAM: We were young when we started. I know you can't hear any Bad Religion, Bad Brains, or Black

Flag in our sound, but those were my favorite bands when I first started thinking about making music myself.

WILL: I remember, you were so excited. You couldn't believe no one had taken that name!

SAM: Yeah, it seems silly now of course, childish even, but the meaning of the name has evolved for me as we've gotten older as people and existed and progressed as a band.

How so?

SAM: Just the idea of flags in general... crosses, logos, signs... We imbue these things with meaning...

WILL: ... and then the meaning gets lost.

SAM: Yeah, so by having a silly name and logo, we kind of avoided that, subverted it somewhat.

DUTCH: Put it this way: If you're not familiar with our music, and you see our name on a flyer, your impression is probably not going to be accurate, and you're not going to expect what we do.

SAM: By the same token you don't want to regularly do something that looks like something else. Eventually you'll be answering to the worst suspicions... Your symbols won't save you.

Speaking of signs, do you guys believe in Astrology?

WILL: No.

DUTCH: Really, Will? You don't think the stars and planets have any bearing on your life?

WILL: No.

A Possible Bad Flag Discography

"Present Tension" b/w "Fragile Fists" 7"
(Vortex Shedding, 1991)

"Bury the Butterflies" b/w "Haunted Halo" 7"
(Vortex Shedding, 1991)

"Get Off the X" b/w "Dogspine" 7"
(Vortex Shedding, 1993)

Born with the Safety Off **(Demo Cassette, 1991)**

Side A:
"Present Tension"
"Fragile Fists"
"Bury the Butterflies"
"Haunted Halo"

Side B:
"Get Off the X"
"Dogspine"
"Wiser" (Coffin Break)
"Spotlight" (Candy Machine)

Rehearsal Bootleg **(Cassette, 1993)**

Side A:
"Get Off the X"
"Present Tension"
"Dogspine"
"Fiery Tale"
"Fragile Fists"
"Fish Fry" (Big Black)
"Walking the King" (Tar)
"Untitled/Entitled"

Side B:
"Where the Day Goes to Die"
"Good God Gone Bad"
"All the Way Down"

"Creeping Tender" b/w "Max Perlich" 7"
(Vortex Shedding, 1994)

Live from the Near-Death Experience (Last show, Fireside Bowl, 1994) (Live Bootleg, Cassette, 1995; CDR, 1997)

"Get Off the X"
"Where the Day Goes to Die"
"Managing the Damage"
"Present Tension"
"Dogspine"
"Weightless Waitlist"
"Max Perlich"
"Bury the Butterflies"
"Gag Box" (Table)
"Fiery Tale"
"Fragile Fists"
"Untitled/Entitled"
Encore:
"Aluminum Siding" (Crackerbash)
"Bob and Cindy" (Johnboy)
"Feral Future"

Man Amok 7" (Touch and Go, 1995)

A: "Where the Day Goes to Die," "Good God Gone Bad"

B: "All the Way Down"

DUTCH: Then why do you go to sleep when this one is facing away from the sun and wake up when it turns back?

WILL: Going to bed when it's nighttime and believing in Astrology are not the same thing.

DUTCH: Whatever.

SAM: No one cops so to their misgivings so easily, but they look good on the side of a bus.

You're about to go on your first tour. Tell me about that.

WILL: We're headed east, to DC, then dipping south to Chapel Hill, Athens . . .

DUTCH: Yeah, we're gonna hit all the hot spots out there, stop off in Alabama to see some friends and visit my family, and then come back home.

You have a reputation for covering some unlikely and very difficult songs, from other underground bands like Johnboy, Table, Butterfly Train, Lungfish, Scratch Acid, and of course, Big Black. How do you choose the songs you cover?

DUTCH: It's just what we like. Sometimes it's someone that influenced us, but it's always a song by a band that we like.

WILL: Yeah, especially where our contemporaries are concerned. We try to put our own spin on all of them, but yeah, it's usually just because we like the song and the band.

SAM: Immolation is the sincerest form of flammability.

BEATING HEARTS: INTERVIEW WITH WILL WILSON, JR. AND SAM SPORTS OF BAD FLAG, 2019

Bad Flag broke up in December of 1994, and they stayed broken up. Not only was a reunion never on the table, now it's not even possible: Their loud and lovable drummer Dutch McNeal was killed in a mass shooting in his hometown in 1999.

The last time I interviewed them in 1993, they were still an active and enthusiastic band. Little did any of us know that they'd break up a year later. They flared up and flared out, but they don't come off as bitter or jaded. If you read that old interview, you may have noticed Sam's penchant for aphorisms. If he seemed a little too ready with a handy quotation, that hasn't changed either.

Numero Group is hereby releasing a full Bad Flag discography. The 3-CD, 5-LP *Dutch* includes remastered versions of all of the band's seven-inches, a set of bonus cover versions, and a proper remaster of their last show at the Fireside Bowl in 1994. In Dutch's honor, I caught up with the remaining members of one of the greatest bands to ever meld minds through music.

Not to start off imprudently, but looking back, some of the quotations in our previous interview seem fake. I've even had a few people tell me as much.

WILL: We liked messing with interviewers, especially Sam or where Sam was concerned. He was the lyricist, so people relate to the words and want to know more. We shielded him to protect that, and sometimes it got out of hand.

SAM: Yeah, at some point, you're not a trickster, you're just a troll.

That's a distinction not a lot of people are going to take the time or effort to make.

WILL: Even still, it becomes something else.

SAM: Gossips never follow up.

That's not exactly fair.

WILL: Well, it cut both ways. There were a lot of misconceptions about us because of the way we dealt with the media, both by being loose and by being closed off. So, we paid for it.

SAM: With your walls up, it's easier for someone to sneak up on you.

I was sorry to hear about Dutch.

WILL: Thank you. Crazy days.

SAM: Yeah, I miss him.

WILL: Even after the band, we all still talked regularly. It's a weird world now.

Really cool of you to name the boxset after him.

WILL: It seemed only right.

SAM: As younger men, I was always off in my head, and Will was all about the business.

WILL: Dutch was the emotional center of the band.

SAM: He was the beating heart.

WILL: He really was.

SAM: It's in tribute to him, of course, but it also has other meanings.

Like?

SAM: It also means dander or trouble, going it alone; looking at the world from a tilted perspective.

WILL: It was also Reagan's nickname, which I believe is where Dutch's parents got it.

SAM: So, even if you didn't know Dutch, it still has meaning.

Speaking of going it alone, tell me about the break-up. You guys planned that ahead of time, right?

SAM: Yeah, it was the difference between it ending and its having an ending.

WILL: It was hard, but it had to happen. We were best friends from middle school to college, and it got to the point where we could either be in the band or continue to be best friends, but there was no way we could be both. We chose to stay best friends.

SAM: It's easier meeting people than it is letting them go.

If you were able to stay together as a band, what do you think about being a band in the current state of the music industry?

WILL: Certain parts of it are great! The ability to distribute your music online is amazing.

And other things?

SAM: You mean social media?

Yeah.

WILL: Ugh. For the bands, it never ends. And as a user, it's like the food in the refrigerator: You keep looking like you're expecting something new to be in there, because you're hungry. That's what social media is. You keep checking, and it's still the same leftovers . . . I'm so glad that stuff wasn't around when we were a band.

SAM: I agree. It seems like a lot to maintain, but complaining about how different things are now betrays a profound and malignant kind of stupid.

WILL: We'd also have to change our name to 'Dad Flag.'

On the back of the Man Amok *seven-inch, it reads, "Bad Flag supports the destruction of mankind." That seems a little extreme, even for you guys.*

WILL: That came from when we were recording. We were at Albini's house.

SAM: Yeah, his girlfriend Heather was really into Norwegian Black Metal, and she had procured this Black Metal magazine from Norway called *Nordic Vision*. In their introduction, it said, '*Nordic Vision* supports the destruction of mankind.' It became a running joke during the sessions. It fits that record thematically as well.

WILL: But we also just thought it was funny, so we put it on the record.

And what about all the cover songs in the box? You guys were known for doing covers live, but where did the new recordings come from?

WILL: That was thanks to Steve. We'd always warm up in the studio with cover songs, and he'd always record them. We hadn't planned on doing anything with them, but then it became a thing. There are even several that survived that we rarely did live.

SAM: One might even say a few classics.

How do you feel about your legacy?

WILL: I feel great about it. We get mentioned in conversations that surprise me, but it's usually in a positive way.

Like what?

WILL: Sometimes we're mentioned in a certain lineage that I'm not sure we were really a part of. It's something much bigger than we were or were intending to be. A lot of it comes from hindsight and the lack of historical context, but it also comes from an intellectual tradition of taking the wrong things seriously.

SAM: And rushing to shelve things in the right category according to questionable or outmoded criteria.

WILL: Yeah, that too.

SAM: They're using wine theory to analyze grapes.

WILL: We had ambitions, and a lot of the specific goals we had in mind we achieved. We started playing live sharing the stage with our heroes, and along the way we shared stages with many more of them, from The Jesus Lizard and Tar that first night to Unwound, Hush Harbor, Engine Kid, A Minor Forest, Candy Machine...

SAM: Lync, Christopher Robin... We played with them on the same night!

WILL: Yeah, it was at the Velvet Elvis in Seattle... Wait, wasn't Candy Machine out there on tour too?

SAM: Yeah, it was us, Hush Harbor, Modest Mouse, Christopher Robin... Candy Machine, and Lync. It was Lync's last show. James Bertram got us on that show.

WILL: Schneider set his drums on fire!

SAM: That's right... What a night! Shout out to Steve and James and Sam and Peter and all those bands.

WILL: Remember that place we stayed in Portland the next night?

SAM: With the big, brown stain on the carpet?

WILL: Yeah, and the—

SAM: Our host tried to assuage our concerns by saying that it wasn't shit, it was blood.

WILL: So, yeah... We played with our heroes, we put out some records, we recorded with Albini... We did a record for Touch and Go.

SAM: But that was never the point.

WILL: Well, no...

SAM: We learned early on that all we could control was the music we were making.

$5 Tuesday 10.11.1994 8pm All Ages.
Velvet Elvis Arts Lounge
107 Occidental Ave S. Seattle, Washington 98104

Even so, you guys are still revered as a band that maintained the underground ethos when everyone around you was groping for the brass ring.

WILL: We appreciate that, but we don't really deserve it. I mean, this band did more than we ever expected. The records, the tours, the songs . . . That's all we wanted.

SAM: It's easy to maintain your integrity when no one is offering to buy it.

WILL: Besides, some people still think we were uptight curmudgeons, old men before our time.

I mean, you guys were straight-edge, vegan teetotalers—still are! What do you say to people who think you're no fun?

SAM: They say that you have to know the difference between a party and a problem. We never even had a party.

Bad Flag *Dutch*

(boxset, Numero Group, 2019)

Disc One: Seven-Inch Discography

"Present Tension"
"Fragile Fists"
"Bury the Butterflies"
"Haunted Halo"
"Get Off the X"
"Dogspine"
"Creeping Tender"
"Max Perlich"
Man Amok:
"Where the Day Goes to Die"
"Good God Gone Bad"
"All the Way Down"

Disc Two: Covers

"Wiser" (Coffin Break)
"Spotlight" (Candy Machine)
"Aluminum Siding" (Crackerbash)
"Bob and Cindy" (Johnboy)
"Speed for Gavin" (A Minor Forest)
"Good Morning, Captain" (Slint)
"Walking the King" (Tar)
"Darjeeling" (Rodan)
"Natural's Not in It" (Gang of Four)
"A Farewell to Kings" (Rush)
"Huck" (.30-06)
"Friend to Friend in Endtime" (Lungfish)
"Windshield" (Engine Kid)
"Fish Fry" (Big Black)

Disc Three: Last Show Live, Fireside Bowl, 1994

"Get Off the X"
"Where the Day Goes to Die"
"Managing the Damage"
"Present Tension"
"Dogspine"
"Weightless Waitlist"
"Max Perlich"
"Bury the Butterflies"
"Gag Box" (Table cover)
"Fiery Tale"
"Fragile Fists"
"Untitled/Entitled"
Encore:
"Aluminum Siding" (Crackerbash cover)
"Bob and Cindy" (Johnboy cover)
"Feral Future"

This story is dedicated to the memory of Sam Jayne.

DUTCH

Subletter

I DON'T KNOW WHAT CHEMICALS WERE STILL coursing through Andy's brain, but I was stuck in the same frantic loop. Dozing off in the passenger seat, I kept waking up when he did, either as he overcorrected the car off the shoulder and into oncoming traffic or the opposite. Gravel and brambles on one side, headlights on bright on the other, and Andy's fevered impulses in between.

We were on our way north from Death Valley, having just spent four days in the desert. Holed up in Lakota country, we'd run out of drugs, and Andy decided it was time to head out about an hour after I'd fallen asleep deep into the last night. If you're not an addict, you don't know what it's like to want something like your life depends on it, and then to get it, over and over again. If you're not an addict, you also don't know exactly why you wake up feeling utterly defeated every morning, why your jaw muscles already ache as much as all the bones in your head.

The sun started burning through the fog just before we hit the 395 to Carson City.

"You want me to drive a while?" I asked after being jerked awake again from the other side of the road. It must've been the extant Ritalin making him so punchy. When you crush it up and snort it, it hits your system much faster, but the effects linger. Without alcohol or weed to temper it, that edge is all you feel. It's like a knife that won't cut. Or a cut that won't bleed.

"Nah, I got it," he said, nodding into the other lane.

I grabbed the wheel and eased us off onto the emergency lane. "Let me give you a break."

I was much more comfortable behind the wheel. Andy immediately called someone to keep himself awake. Why he didn't do that while driving or why he didn't just talk to me, I don't know. This trip was going to stay as uncomfortable as it had been since we left Seattle after our cover had been blown the week before. Maybe he was still mad at me.

The house on Stone Way in Seattle's Wallingford neighborhood had a mail slot in the door. The mail hit the hardwood with an audible *woosh* or a dull thud. Most of what landed in the floor was either junk—postcards from direct mail campaigns, colorful but disappointing sale papers, or envelopes with plastic windows claiming urgency—or leftovers for previous tenants.

In the last few months that I lived there, postcards addressed to a Brandon Strand showed up with ever-increasing frequency. I assumed they were for a previous tenant or another subletter. Someone I wasn't familiar with, but who I wished I could locate. The messages on the cards were wise and intriguing and sometimes included requests, like kites in prison.

My father cheated on my mom once. I only know because I caught him. It was just like when I walked in on my college roommate while he was masturbating. We exchanged a knowing, uncomfortable look, and neither of us ever said another word about it.

Shared secret knowledge like that undermines bold statements. If dad ever started to preach the importance of fidelity, I could hear his resolve creak.

His was a world of plastic five-gallon buckets, aluminum ladders, tool belts, and drooling paint cans. In his way, he tried to teach me what he'd learned. He taught me to tilt my coffee cup only when the driver drinks unless you want a lap full of hot coffee, that California air is charged with the silent power of the possible, and that slowing down is not the same as stopping.

The first postcard I got for Mr. Strand reminded me of my dad's cryptic wisdom.

If you wish to walk
With feet 'pon wings
Then, my son,
You must learn a few things

You must be patient
And make moves ever slow
Like those of the ancient
Lands down below

You must tickle hot buttons
With cool, cool fingers
You must suspend suspicion
Even as it lingers

You must be confident
And mind the edges
You must lie on the line
All the bets the other hedges

Strand's correspondent was much more poetic than ol' dad, but I got the impression that they'd agree on a great many things.

One day, just as I was picking up the mail, the knuckles of a fist met my door in an urgent manner. Sometimes the mailperson would have to knock to leave me a package that was too big for the slot, but this was not their knock. I looked through the peephole to see two men in suits and dark glasses who looked like television cops. I hesitated to answer but went ahead because I figured they'd already heard me walking on the hardwood.

"Mr. Strand?" one of them said as I opened the door.

"Um, no, I don't know any Strand, though I have gotten some of his mail here."

"We're going to need to ask you some questions. May we come in?"

"Sure."

Their crossing the threshold of my front door felt like a stab. Their presence at the door at all felt threatening, but their being in my home made it sharp.

You've been in a situation that, once given the

possibility of escape, no amount of money would keep you in, right? This was just such a situation.

"Can we see some identification?" the other one asked.

"Can I?" I shot back. They both unfolded their wallets to flash their badges and FBI cards. Total cliché move. Television cops. Slightly amused but satisfied, I pulled my driver's license out of my wallet and handed it over.

Once they too were satisfied that I wasn't Strand, they finally left. The breathable air returned to the room.

About a half an hour later, Andy plopped down on my couch with a plastic shopping bag full of shit. His timing was suspiciously convenient, especially considering that he claimed to know Strand.

"Yeah, etch-a-sketcher," he said looking at the latest postcard. "He comes and goes, but he mostly goes." The coffee table's front edge was scored from breaking pills in half. He laid the card down on a light dusting of powder and started unloading the bag. "Party favors, *por favor*."

"The police *just* left my house," I protested.

"Well, you didn't exactly clean up before their visit," he gestured at the table with a bag full of pills.

"They don't really call ahead." Ignoring me, he started crushing up a few pills on the table. "You're not doing that here."

"I am. The rule is '*do* drugs don't *have* drugs.' Right now, we have drugs, so we must do them."

Though his logic was impenetrable, I was already on edge from my previous visitors. The last thing I wanted to entertain was a pill-addled Andy on my couch. I lunged for the bag, but Andy snatched it away. I tried again, but he was too fast. He must've partaken on the way over. I tried one more time, but he pulled a shiny, silver revolver out of his waistband and leveled it at my left eye. "Sit down."

Suddenly sober, I eased myself into the chair across from him.

"We have a very good problem here. Normally, our struggle is finding this shit. Right now, we have enough for a week—a week and a half if I shoot your annoying ass. You and I are about to find the bottom of this bag and three more just like it."

"Well, we can't do it here."

"No, we can't. This amount of fun requires the proper setting. We're going to the desert, Hunter S. Thompson–style."

"What desert?"

"Death Valley. I know a Lakota girl down near Darwin, and I have an open invitation. She'll tolerate you."

As he leaned over to snort up a choppy line, I lunged at him again. I grabbed his gun-arm just as he squeezed off three quick rounds. A scream and a body-dropping thud echoed through the ceiling. I hadn't thought my upstairs neighbor was home. After a slow, silent moment, we scooped up the contraband and stumbled out to Andy's car. In the passenger seat, I pulled on a stocking cap and struggled into a hooded sweatshirt I managed to grab off the back of the door on our way out.

That day reminded me of the aftermath of my last bad breakup. It was an end that felt like the beginning of something else. The pick-up parties started only days after she left. Friends I hadn't seen in months, some in years, would come by with beer, pizza, Patrón. Friends I didn't even know I had would catch wind of a gathering and come by as well. The stripper from two doors down, Carrie, stage name "Candy," would come over just as other people started leaving. We'd drink, do lines off the counter in the kitchen, and kick a soccer ball around the living room until the sun beckoned her back home. She always had a rolled-up one-hundred-dollar bill at the ready. She'd stay all night, but she would never go to bed with me.

Those nights were fun, but it was the kind of fun that held gloom right over your head, ready to drop at any second.

The Lakota girl's invitation was more of a debt. Andy owed her a lot of money. Somehow, I already knew this. Maybe because he owed me money, and he always tried to pay me in drugs. He valued the drugs more, so he thought everyone else should, too. Andy's lines of logic were impervious to his own scrutiny. He never thought twice about whether the way he felt about something was the way others should feel about it, regardless of how they acted or told him they felt. Call it a blind spot or a superpower. It wouldn't matter to Andy.

The time we spent in Darwin isn't worth recounting here. It reminded me of the time our neighbor was trying to sell some stereo components. A receiver, a dual-tape deck, and an equalizer. When he couldn't get what he wanted for them, he took them out into the field behind our houses and smashed them with a hammer.

After Andy finally dozed off, I started eying places to ditch him along Route 395. I'd long had enough of his manipulations, and I knew I needed to make a move before he woke up.

Just before the California–Nevada line, I spotted a reservoir and pulled over. I grabbed the gun, his wallet, and secured his seatbelt. I hopped out and let the Jeep roll down a boat ramp just off the road. I threw the gun as far as I could out into the water.

As the car gurgled into Topaz Lake, I checked Andy's wallet. Along with a lousy 17 dollars in ones, there was an Oregon driver's license with Andy's ugly mug next to the name Brandon Strand.

I won't tell you which way I headed when I left Andy in that lake, or what I was driving. I might send you a postcard, but don't expect a return address.

Hayseed, Inc.

Philosophy of science is as useful to scientists as ornithology is to birds.

— RICHARD P. FEYNMAN —

A woman needs a man like a fish needs a bicycle.

— IRINA DUNN —

"DO I LOOK INVITING?"

She caught his stare before he did. She had the morning kickstart: tall cup of coffee and a cigarette. Todd only noticed that because he craved the same. The sound of her Zippo hit his consciousness like a rock tossed into calm waters. Its metallic slap and the smell of tobacco burning was the only reason he turned and let her presence on the steps behind him fill his vision in rippling concentric circles. He didn't even know if his answer to her auspicious question made it past his lips.

She took a long drag off the Camel and put her hand on her hip.

"Do I look inviting?"

The fall rain had let up just long enough for the cold to reassert its bite. The sky was slate gray and the only sentiment on the faces he passed on the way to his stop was that of wanting to stay home in warm beds.

"Do I look inviting?"

She looked at him hard, as if she was trying to check his pulse. Just then the bus lumbered up to the curb. He swore he answered her, but before he could she said, "Well, you're invited," and disappeared up the steps and onto the bus.

He thought he followed her, but he didn't really remember. The last few weeks had been just like the few weeks before that, and the few before that: just weeks that passed seemingly without event. Hearing that Zippo click was the prelude to the first truly noticeable occurrence since he'd moved here nearly three months earlier.

We needed people to abandon their interests and pursue a goal we could profit from. Something frivolous yet lucrative. Our first few attempts were failures. Politics are needlessly convoluted, and compromise is around every corner. The money comes but only after great efforts. Sports are similar, but there is something physical, tangible, operational about them that politics lacks. We needed our rubes to be able to fake their talents as much as possible. Finally, we arrived at gambling.

There are many established and acceptable ways to gamble representing an array of mysterious skills. It's much easier to fake something if no one really understands how it works. And people are drawn to the unknown. Mystery loves company.

The roulette wheel, the rolling of craps, slots of all stripes: the unknown abounds. But even in the various rule-based and quantifiable card games there is uncertainty. There are professionals who are untouchable at these games, yet there is always a chance someone new will sit at the table and win. The wildcard. That's us.

We have made sure there was a possibility of a wildcard since the great games began. The hayseed's chance in hell of unearned riches and rewards. In politics, it's possible. In sports, less so. In gambling, we've been holding that door open the whole time. People seem less averse to risk in the face of a beginner winning. People who'd normally be fairly staunch become slightly gullible. The possibility of wealth wears down their logic just enough. So, as we've figured out the angles to shoot, we've been expanding the operation.

As long as there's the impression that anyone can do it, we are in business. So, we sponsor the hayseeds. We fund them, and we rig the games in their favor. That Blumstein kid from New Jersey in 2017 beat out 120,995 players for the win at the final table. Not good odds for him going in, but those are the numbers we like. We found the kid online, playing poker.

Blumstein's final hand of an ace of hearts and a two of diamonds ended up being stronger than that of Ott, who went all-in with an ace of diamonds and an eight of diamonds. The community cards were a jack of spades, a six of spades, a five of hearts, a seven of hearts and a two of hearts. It was the last card that prompted Blumstein's supporters to erupt. "I'm really happy with the result," he said, "really happy with the deuce because I was playing good." One of those twos was us.

"I'm really happy about how I played tonight," the

well-trained Blumstein said. "This is just one poker tournament. It takes variance and luck and playing your best, and all those things came together, and I'm happy to be the winner." He'd have to be. As an accountant, he knows what $8.1 million does to a budget. And so do we. We've even gone global. That Hacham guy in Australia in 2005? He was one of ours too.

In business terms, we're a consulting firm. If a business needs droves of gullible players or pawns with moderate-to-full bank accounts, we can make it happen for one weekend player and make it believable to the rest. Maybe their company needs one success story a year to keep the masses' money moving their way. Maybe they need One Big Win and the scam is set for history. Either way, we can make it happen.

We are the equalizers. We democratize the illusory stakes of any game. If anyone can do it, you can do it too. We make you believe in yourself.

Todd walked into English and took the first available seat in the back. Two desks over sat the girl from the steps. She was decidedly pale, with short, jet-black hair thoughtfully piled on her head. He pulled out his notebook and tried as subtly as possible to get her attention. She toyed with the tiny troll on the end of her pencil and stared out the window. When the bell rang and she got up to go, he followed.

Her friends surrounded her as they all poured outside. She introduced him to them as if they'd been talking the whole time. They talked and laughed and hung out from then on. They were holding hands by the time everyone decided to head home.

The way most of us see the world relies on a belief that all the mysteries of life are eventually knowable. Many of our realities hinge on the fact that all will one day be revealed, or that we'll at least get a glimpse at what's really going on as we move through this life, that it's not all just some matrix of coincidences. Our being is bound by time and space, and unrooting anything from that ground requires knowledge from somewhere else.

In one of our toughest and oddest cases, a dad hired us to make his son popular at school. This required a finesse no other case had had before or since. As much as we prefer to work directly with clients and not a proxy, especially a parent, dad paid our full fee and expenses—and the challenge was just too enticing.

High school operates according to steadfast yet unknowable rules. Some famous scientist once claimed that God had a big book of the rules of the Universe, and that every once in a while, scientists were given a glimpse. The rules of high school are in a much bigger book that is much more rarely glimpsed.

There's a group in every high school that seems to have a copy of that book. Remember them? They got away with things that confounded not only you and your friends but your enemies as well. At Elbo High, they were a small, tight group who wielded seemingly unlimited power from the four corners of the school: Norrie, Michael, Craig, and Sean. They were popular but not in the popular clique. They were into sports but not jocks on any of the school's teams. They were smart but not nerds in any after-school clubs. They were bad boys who shifted between groups and mingled with their members at will. They were untouchable. Every guy wanted to be one of them. Every girl wanted to be with one of them.

The four of them were never seen all together at once. Two of them would be sitting at the same lunch table on Tuesday. Three of them convened at Larry's Barbecue for burgers after school. Two others hung out in the parking lot after the Friday night football game. Students liked to imagine that the ones seen were keeping up their public face while the hidden ones made backroom deals with Principal Carter, Vice Principal Thompson, and Coach Casey for more power of the school.

Knowing all of this, we infiltrated their group and installed Todd.

Of course, he hoped this would happen. Of course, he didn't actually think it would. But here he was, walking home hand in hand with a beautiful girl that he'd just lucked upon during a regular school day.

Dreams don't come from staring at screens.
Dreams come from doing things.

He woke up to her singing in the shower. He'd fallen asleep on her couch. He got up and checked the time: 7:21.

Maybe just a little bit
May be just enough.

"What is that you're singing?" Todd yelled through the door.

"Bad Flag," she yelled back. "Why don't you come in here, and I'll sing you more while you wash my back?"

It's like rain you can hear falling,
But can't feel on your skin.
A thirst you drink to death,
But can't get your fill.

Her teeth were her whitest aspect, barely beating out her skin. She had dyed the bobbed mop on her head a shade darker than midnight. Normally, this was exacerbated by the fact that she wore nothing but black. She looked like a photocopy of a photocopy of a photocopy, or a black and white CRT with the contrast turned all the way up.

You drink like you're driving,
And you drive like you're drunk.
Another steering-wheel sing-along
With the body in the trunk.

Naked, though, her teeth were the polar bear in her blizzard-snowstorm skin, while her while her pale pink lips and black hair offered the only contrast. The white tile of the bathroom only added to the scene.

Brake-light hesitation,
And gunpoint inspiration.

Maybe just a little bit
May be just enough.

She was like a cross between Molly Bloom and Molly Millions. She was never without her dark glasses and was given to soliloquies of great length and thick description. He thought of her as a character being played by someone else. They drank coffee and talked in the waning evening sun. Her *Philosophy of Science* textbook lie open on the counter.

"I think Thomas Kuhn was just a serial monogamist," she said, stirring sugar into her coffee. "See, he's with this one girl for a while for a period of 'normal science.' Then, he meets some other girl, bam! Revolution! Paradigm shift! Then he settles down with his new skeez for another period of 'normal science.'" She took a self-congratulating first sip of her coffee. "Say what you want, but it makes more sense when applied to romantic relationships than it does to the progression of actual science."

"Interesting... So, how would your theory cast Feyerabend?" he asked, playing along.

"Total player. That guy got around," she answered without hesitation.

"I guess that makes sense. He was a breadth-man after all," he smiled slyly. "How about Lakatos?"

"He's a little more tricky, possibly even gay."

"Oh, you know sexual preference doesn't have any bearing on promiscuity!" he protested.

"No, but it does make it more difficult to fit him into my theoretical framework."

In many ways, Todd was a model client for us. Nondescript. Nobody. Anybody. If we could make him popular, anyone in the school could be popular.

Successful disruption of a system of this sort requires not only upsetting its natural order but also rearranging its rules. In order to get Todd into this group, we had to make it their idea. Since we owned none of the four, we had to bring in something they wanted and give it to Todd, so they'd want him. To that end we brought in the one thing that can wreck any stable high school social structure. We brought in The Girl.

The Girl, let's call her Heather, was selected through a rigorous casting process. This role required a believable blend of blemish and polish. She had to be flawed yet perfect, petty yet mature, rugged yet feminine. She had to dupe the four kings of the school by first duping Todd. She had to be desirable to the four kings but also accessible to Todd. She had to have reason to dis royalty but also be convincingly into Todd. Casting also required information about Todd's preferences that Todd might not have been forthcoming about. So, with his dad's signature on a waiver, we hacked his phone and computer, checked his browser history, vetted his friends on social media, and perused his journal. From this dataset, we created a profile. After auditioning hundreds of girls, we got it down to three. Once we explained what the role really was—and why it paid so well—two of them demurred, leaving only Heather.

Heather was pretty but in an everyday kind of way. Model material in some neighborhoods but the girl next door in others. Her beauty was disarming, not flashy. She was also smart. This was key, not only because Todd was kind of a nerd but also because she had to learn a lot of details to play the part. We needed her to play this thing naturally. We needed her to know things but not act like she knew them. We needed to control her, but we needed her to be autonomous. We needed her to want the same things we did. We needed her to be motivated toward our goals. Sure, the money was supposed to do that, but everyone knows intrinsic rewards always win. We really needed her to actually like Todd.

Heather enrolled at Elbo High a few weeks into the fall semester. All of the social cliques were well established by then. She was an outsider, but being the New Girl would be a boon not a burden. Where others might have accepted their place by the wayside, Heather's looks and swagger leveraged that imbalance, vaulting her to the envied echelons of the social structure of the school. Our plan dictated that on her way up, she take Todd with her.

And it almost worked.

"Do I look inviting?" said the girl on the steps on the television.

"What's this?" he yelled to Heather in the next room. She walked in and looked, plopping down on the couch next to him, one leg draped over his lap.

"This is that pilot I told you about, the one I was in that we filmed a few months ago. I didn't know they were going to actually air it."

"Say, did you do any method acting on the streets in preparation for this role?" he asked, looking her way.

"Yeah, I'd stand on the steps of the building I used to live in, and do the lines to strangers waiting for the bus, why?"

"Huh," he said, looking her over. "No reason."

She was only wearing white underwear and a t-shirt, and when she caught him staring at her thighs instead of the TV, she grinned and added, "Well, do I look inviting?"

As he started to kiss her, sliding his hand slowly up her thigh, she stopped him.

"Hey, tell me something..."

"What?" he said, continuing his advance.

"How did you know I was into you that first day at school?"

"I caught a vibe," he said, kissing her neck, sliding his fingers under the hem of her underwear. "Well, that and you were oblivious to Craig and Sean, and they were being their usually charming selves. I put two and two together and hoped for the best."

Our plan ultimately failed. Once we got them together, Todd no longer cared about being popular. Once Heather inevitably broke his heart, he didn't care about anything. He didn't kill himself, but he tried. His dad sued our firm. We settled before it went to trial.

Afterward we went back to our bread and butter: gambling. We even gave politics another go. The variables there are easier to control, and the stakes are much, much lower.

POST-INTELLIGENCE

COLD OPEN

EXT. OFFICE PARK PARKING LOT—MORNING

FADE IN:

Theater Editor, MAX [mid-30s, paunchy white dude], is smoking beside his car, a souped-up 1968 GTO. He's wearing jeans, dress shoes, and a button-up with no tie.

A new hybrid pulls into a spot marked EDITOR, ARTS AND ENTERTAINMENT. A sticker on the back window shows a stick-figure family with the son in a graduation cap and a diploma in hand, and the mom holding a cocktail. DORA, a Latina in her late 50s, steps out wearing high heels and jeans with a simple blue blazer.

Max turns as Dora emerges smiling.

MAX (sarcastically): I'm sorry, ma'am, you can't park here. This spot is reserved for the Arts and Entertainment Editor.

DORA: That's still me for one more day. (*nodding at Max's cigarette*) Those things'll kill you, you know.

Max points his cigarette at Dora's Prius.

MAX: When did you get that?

DORA: After the divorce. I went through the requisite dark period, and then I had an epiphany. I'm not going to be the problem anymore. That's his thing. I am a part of the solution now.

Max drops his cigarette butt and stamps it out.

MAX: High profile.

EXT. OFFICE PARK—DAY

Dora and Max are walking from their cars into a boxy blue 10-story office building. A giant globe is rotating on top, encircled by the words POST-INTELLIGENCER. A 6-foot eagle sits on top of that. Modern yet outdated, the building looks somehow both old and new at the same time.

MAX: I hate when someone recommends a TV show to you—

DORA: —Me too—

MAX: No, I don't necessarily hate the recommendation. I hate trying to figure out why.

DORA: Yeah. How was your vacation, by the way?

MAX: I'm getting to that—Because then when you're watching, you're filtering it through everything you know the recommender knows about you and your viewing tastes. So, you're watching through this weird social lens.

DORA: It's like life in the 21st century.

MAX: How's that?

DORA: We live "as if" there's an audience always watching, so everything is filtered similarly to the way you just described.

MAX: Huh.

They walk for a beat.

DORA: And your vacation?
MAX: The show was about a ragtag bunch of people in the office setting of a failing paper business.
DORA: Sounds familiar. Wasn't that show canceled years ago?

Max shrugs.

MAX: Vacation didn't happen. We stayed home and binge-watched TV shows, hence the recommendation.

They reach the office building just as a guy in a giant banana suit walks in. Max holds the door open for him/it. They go in after.

MAX (cont'd): What's with the banana?
DORA: It's for the meeting.
MAX: Ah. The meeting. [. . .] You know what they say about a banana in the first act!

INT. OFFICE BUILDING—DAY

Dora and Max are walking down the hall by giant newspaper printers, which are being disassembled for sale to a Canadian printing house.

MAX: He's late every day, so I made him stay late in the evening for every minute he's late in the morning. That didn't work so I started docking his pay. He's still late—sometimes later!
DORA: That's load-bearing guilt!

MAX: What?

DORA: You removed the one thing that gave you any leverage in the situation: his guilt for being late. He doesn't care about the consequences, and now he doesn't feel guilty for being late.

MAX: Oh.

DORA: It's classic Freakonomics.

MAX (incredulously): Of course!

DORA: Hey, remember Madison?

MAX: No.

DORA: She used to be on the delivery staff, but she's also a blogger. She sucked at deliveries, so I brought her in to help you with social media.

MAX: A blogger?

DORA: Yeah, she has a master's degree in Information Science.

Max looks confused.

DORA (cont'd): You know, libraries, indexing, and archiving, but she started out making 'zines.

MAX: 'Zines?

DORA: Yeah, you know—little homemade photocopied magazines. [. . .] The point is that she's quite familiar with print as well as digital media.

MAX: The point is that I don't think you're speaking English. Are these 'zines anything like scrap-booking?

DORA: Kind of.

MAX: My mom does that stuff, and it's not really like making a newspaper. By the way, that reminds me:

Can you get me a few cans of that spray sticky-stuff before we give it to the Canadians?

DORA: We're not giving it to the Canadians. We sold all of our printing equipment to them.

MAX: Why do they want it anyway? Doesn't America outsource all of its printing to them in the first place?

DORA: You just answered your own question.

MAX: So they want a monopoly on 15th-century media technology?

DORA: Something like that. [. . .] You don't think there's a Canadian samizdat outfit that needs a printer in Seattle?

MAX: Would that be a 'zine or a blog?

DORA: I'm never going to get to retire.

[MAIN TITLES]

ACT ONE

INT. DORA'S OFFICE—DAY

Dora and Max are having coffee and looking through the blinds as they prepare to join the rest of the STAFF slowly assembling in the conference room outside her office. The staff (including MADISON [25, glasses and tattoos: nerdy, Suicide Girl] and PHIL [22, Nazi-youth haircut, preppy punk]) are gathered around a conference table listening to a presentation by MITCHELL [28, MBA head-to-toe], the new Online Marketing Director.

Max motions out the window at the table to Madison.

MAX: Who's the new girl?

Dora looks over Max's shoulder out the window at the STAFF.

DORA: That's Madison. I just told you about her.
MAX: She's the 'zine-slogger.
DORA: Something like that.

Max takes a closer look.

MAX: High profile. I like her.
DORA: You just like her for her ta-tas and ya-yas.
MAX: No, I just like a tire with some tread left on it. (*picking up his phone*) Is she on Instagram?
DORA: You wouldn't prefer one with a little more mileage?

Max side-eyes Dora.

MAX: Are we still talking about journalism?
DORA: I'm not sure, but if not, we really should watch what we say.
MAX: Oh, c'mon... My office is our "safe space." She is on Instagram. Lots of followers too—
DORA: —It's not your office yet, it's still mine for the day.

MAX (looking up from his phone): This day has come before. If you'd ever stop talking about it and actually retire, it would be mine.

Dora gets up to put on her blazer to join the meeting.

DORA (gesturing dramatically): Until that day—

Max gets up too. And holds his phone up to Dora.

MAX: Look at this.

DORA: She has more followers than—

MAX: —Than the *P.I.*

DORA: Solid hire, Dora! We'll have to get her together with Mitchell.

Dora opens the office door to the meeting in progress, where Online Marketing Director Mitchell is addressing the STAFF. The Banana is standing next to him.

MITCHELL (o.s.): No longer will this revolution happen without us, because believe me, it's happening!

INT. CONFERENCE ROOM—DAY

Dora and Max step out of Dora's office into the meeting in progress.

MITCHELL (motioning to the banana): What we're doing here at the *Post-Intelligencer* is like peeling a banana: By ditching our paper version and going all-digital, we're peeling away the skin to get to the good part!

He dramatically peels the Banana.

After a beat, there is scattered applause and snickers from the STAFF. Everyone except Madison manages not to roll their eyes.

INT. BULLPEN—DAY

Madison and Phil are sitting at their desks, morning coffees in hand. Dora is making the rounds before returning to work.

DORA: Was that a little over-the-top or what?
MADISON: Beyond the peel, maybe?

They laugh. Phil does not.

DORA (laughing): Yes!
PHIL: I call that line the "Wedge Limit."
MADISON: "Minimum Wedge."
PHIL: What?
MADISON: "Minimum Wedge" is a better name for it.
DORA: A better name for what?!
MADISON: Phil has this idea—
PHIL: —It's the extent to which you can push something iffy—a pun, a play on words, a marketing idea, a shocking advert—without turning off your audience. At a certain point, you can get people's attention, get them to reassess their attitudes, make them laugh, but past that point, you turn

them off—drive a wedge between your message and your audience—

MADISON: "Minimum Wedge."

PHIL: "Wedge Limit"!

Dora rolls her eyes and grabs her stuff to leave.

DORA: Found it. (*addressing Madison*) Come see me when you get a minute.

Madison and Phil stay at their desks. He's still slightly annoyed. She's checking her phone.

PHIL: What do you think that was about?

MADISON: Oh, man!

PHIL (half interested): What?

MADISON: Twitter is saying that Bob Stephenson just died!

PHIL: No way! The dildo-luggage guy from *Fight Club*?!

MADISON: Yeah, it's blowing up. I have to take this to Dora.

Madison gets up, her attention still on her phone.

PHIL: Well, you're supposed to go see her anyway.

He slowly gets up to follow her.

INT. DORA'S OFFICE—DAY

CUT TO:

Max and Dora are talking after the meeting. Max is leaned way back in Dora's desk chair, his feet propped up on her desk. Dora stands just inside the door. A few staffers new and old are outside the office talking, slowly breaking off to their desks.

MAX: When did "vinyl" become a singular noun?

DORA: What do you mean?

MAX: The kids say that they "bought a vinyl" or "listened to their vinyls." What happened to "records" or "albums" or "having something on vinyl"?

DORA: Who cares? You're just jealous.

MAX: Jealous of what, sounding stupid?

Madison, phone in hand, bursts into the office with Phil in tow.

MAX (addressing Madison): Okay, what does this fart smell like?

EVERYONE (stepping back): What?!?

MAX: My father used to say, "what does this fart smell like? If it smells like ass, then this too shall pass. If it smells like shit, there's more to deal with."

Everyone exchanges uneasy glances. Phil sniffs the air.

MAX (cont'd): It's a metaphor. So, what does this fart smell like?

PHIL: Well, it's more of an idiom—

MAX: Whatever! What is it!?

MADISON: Twitter is blowing up with claims that Bob Stephenson just died.

MAX AND DORA (in unison): No way!

They all pick up their phones or turn to Dora's computer screen. Dora steps around the desk as Max swings his feet down to the floor and frantically types in a search.

MAX: That cannot be true. I loved him in *Fight Club*.

MADISON: I haven't found where it originated yet, but we need to say something.

MAX: Call me silly, but—

EVERYONE (in unison): Silly butt!

MAX: Silly-comma-but—

EVERYONE: Silly comma butt!

MAX (undeterred): We need to verify this first, of course.

DORA: Madison, you take lead, but take Phil, and get the rest of the web team to dig too.

Madison nods and turns to go, thumbing her phone. Phil turns to follow.

MAX: High profile . . . The first crisis of our new era!

DORA: This one smells like shit.

PHIL (over his shoulder as he walks out): Actually, it smells like cabbage.

DORA (addressing Madison): Hang back for a minute.

MADISON: What's up?

DORA: I don't know how I hadn't noticed before, but I saw your Instagram page.

MADISON: Oh, god! That picture! I was planning on deleting it. I just haven't—

DORA: —It's not about that, though now I have to see this picture!

Madison's face reddens.

DORA (cont'd): No, I saw what a platform you have on there, and I want you to sit down with Mitchell.

MADISON: The marketing banana guy?

DORA: Yeah. He knows how to leverage such things. Meet with him, show him your account—with or without the questionable picture—and report back to me.

INT. CONFERENCE ROOM—DAY

A few staffers are sitting around the conference room table brainstorming an article. They're going over notes.

FEMALE INTERN (off her notebook page): These are the ones I got last night.

MALE INTERN (reading her notes): Neil Patrick Harris, Jennifer Love Hewitt, Robert Sean Leonard, Seann William Scott, Philip Seymour Hoffman, Philip Michael Thomas, John Michael Higgins...

Dora walks up, her arms full of papers.

DORA: What's this?

FEMALE INTERN: Jennifer Jason Leigh, Hillary Rodham Clinton, Melissa Joan Hart,

Malcolm-Jamal Warner, Helena Bonham Carter…

MALE INTERN: "Malcolm-Jamal Warner" isn't a three-name name, it's a hyphenated name. They don't count.

He turns to Dora.

MALE INTERN (cont'd): It's a test listicle—Chloë Grace Moretz—

DORA: Ah…What's the topic?

MALE INTERN: Famous people who go by three names. Sarah Michelle Gellar. Wait, who's John Michael Higgins?

DORA: Like on *The Larry Sanders Show*?

FEMALE INTERN: He was the Seize-the-Day professor on *Community* and Elaine's bald boyfriend on *Seinfeld*. He also went to Amherst with David Foster Wallace!

MALE INTERN: There's another one! Where did you get these?

He turns to Dora.

MALE INTERN (cont'd): The who?

DORA: *The Larry Sanders Show*. On the third episode of season two, the guests were Edward James Olmos and Mary Stuart Masterson, and the musical guest was John Wesley Harding.

MALE INTERN: Yes! Add those to the list!

The Female Intern scribbles them in her notebook.

MALE INTERN (cont'd): Wait. How do you know that? Have you two been hanging out?

FEMALE INTERN (still writing): Mark David Chapman!

Everyone turns to look at the Female Intern.

DORA: No, I just re-watched it last night.

FEMALE INTERN: What?! He's famous. [. . .] You all know who he is!

DORA: What's the story? A listicle is more than just a list.

MALE INTERN: That's why it's a test—

FEMALE INTERN: It's a testicle!

Giving the Interns a blank stare, Dora gathers her papers to go to her office.

Then she pauses.

DORA: Rachael Leigh Cook, Evan Rachel Wood, Paul Thomas Anderson, Philip Baker Hall, John Paul Jones, John Wilkes Booth, James Ward Byrkit, David Wilson Barnes, Caleb Landry Jones, Thomas Haden Church, Mary Elizabeth Winstead [. . .] Peter Mark Richman, Stephen McKinley Henderson, Tim Blake Nelson, Brian Austin Green, John Carroll Lynch, David Robert Mitchell, David Alan Grier, Sarah Jessica Parker, Mary-Louise Parker, Harriet Sansom Harris, Sarah Michelle Gellar, Mary Elizabeth Ellis, Jamie Lee Curtis, Julia Louis-Dreyfus. [. . .] You guys need to come up with something that's going to get A&E on the front of the website—

MALE INTERN: We have Sarah Michelle Gellar already, and Mary-Louise Parker and Julia Louis-Dreyfus are hyphenated—

DORA: —some entertainment news that everyone will care about.

FEMALE INTERN: There's the Bob Stephenson thing, if it turns out to be true, but—

DORA: I think Madison has that covered. [. . .] You all need to do some real reporting.

Dora turns to the Female Intern as she walks away.

DORA (cont'd): And maybe switch to decaf.

INT. OFFICE—DAY

Dora is sitting on her couch. Phil is sitting adjacent in the other chair. Her desk chair is empty.

DORA: It's a problem of scale.

PHIL: How do you mean?

DORA: You can have a passion for gardening and not have one for farming. [. . .] I know you've been at the *P.I.* for a while, but you've only been with us for a few weeks. With all of the changes, there could be a real opportunity for you here.

PHIL: Yeah, I like designing things and I even like writing, but that's not really what I do here.

DORA: Give it another week and see how you fill, fill (*"feel, Phil"—She pronounces the last two words exactly the same*).

PHIL: What?

DORA: Just give it a little more time. Please.

PHIL: Okay.

Phil picks up a picture off Dora's desk and turns it around. It's her and her son posing proudly at his high-school graduation.

PHIL (cont'd): Is this your son?

DORA: Yeah.

PHIL: He's quite handsome.

DORA: Like his father.

PHIL: He's also quite . . . not brown—compared to you, I mean.

DORA: Yeah, he's more chip than salsa.

Phil puts the photo back down, stands, and steps toward the door to leave.

DORA (cont'd): Think of something you want to do for the site that seems under-represented to you, something you're into that no one else here is, but also something with an audience. Remember your, what was it? "Wedge Limit"?

Phil steps back toward her desk.

PHIL: I have conceded to the name "Minimum Wedge."

DORA: Think of something you know about, want to know more about, and that you want to tell people about. That's the sweet spot of entertainment journalism.

Phil smiles.

INT. BULLPEN—DAY

Madison grabs her notebook, jumps up from her desk, and runs through the bullpen to Dora's office.

INT. DORA'S OFFICE—DAY

Max is on the computer. Dora is sitting in the corner thoughtfully working her teeth with a toothpick. They look up as Madison barges in.

MADISON: I found the source.

MAX: The source of that smell?

MADISON: The source of the Bob Stephenson death rumors.

MAX: Oh. Who is it?

MADISON: A comedian named Brendon Walsh.

MAX: The guy from 90210? Does he know Bob Stephenson?

MADISON: No. Neither. He's a comedian.

MAX: Oh, well, make sure.

MADISON: Make sure what, that he's funny or that he's not from 90125?

MAX: 90210! Make sure he doesn't actually know Bob or his people!

MADISON: He doesn't. It was a joke.

MAX: Make sure.

MADISON: Dora!?

DORA: You know how this works, Madison. Get confirmation or refutation just like always.

Madison walks out. Dora's intercom buzzes.

DORA (cont'd): Yes?

INTERCOM: Dora, you and Max are requested on the 10th floor.

Dora and Max both cross their fingers.

DORA: For Tom?

INTERCOM: No. For Trace.

They hang their heads.

DORA (defeated): Okay. We'll be right there.

INT. MITCHELL'S OFFICE—DAY

Mitchell is at his desk. Madison is sitting in a chair across from him. They are staring silently at each other.

INT. 10TH FLOOR—DAY

The elevator doors open to a museum-like space: high ceilings, strange art, sparse furnishings, etc. Oppressively loud ambient music fills the air. One lone desk sits back toward the far wall in front of what looks like an aquarium but isn't.

TRACE [indeterminate age and gender] is floating in the space behind the glass wearing a sleek, all-black body suit, VR goggles and gloves connected to a bundle of multicolored wires and cables. They move in flows and fits as he does.

Dora and Max slowly step into the room.

DORA: Hello? Trace? [...] You wanted to see us?
TRACE: Yes. One moment.

He twists and spasms a few more times before he is finally still. Lying back suspended in the space, he folds his hands across his chest.

TRACE (cont'd): Dora, Maximillian, Welcome. Can I get you anything?
MAX: Sure, I'll take a—

Dora elbows him sharply in the ribs.

DORA: No, thank you. What's this about?
TRACE: Bob Stephenson.
DORA: I think we have that under control.
TRACE: You need to drop the story.
DORA: I'm sorry?
TRACE: Drop it. Let someone else lead. Follow up if you have to. Then don't ever mention him again. [...] He wasn't nice and he could be different from useful.
DORA: Okay.
MAX: But...

Trace's body has been slowly drifting, rotating, and now he is upright, still floating in the space behind the glass. His eyes are still shielded behind the VR rig, but he is clearly addressing Max directly.

TRACE: Maximillian, need I remind you that there are things in this life larger than you are? Seattle may be a city of relative size, but it still sits squarely in a small world, a world connected in every possible way to every other. [. . .] Drop the story, or I'll be forced to show you just how connected it is.

ACT TWO

INT. MITCHELL'S OFFICE—DAY

Mitchell is at his desk. Madison is sitting in a chair across from him. They are staring silently at each other.

INT. DORA'S OFFICE—DAY

Max is sitting at Dora's desk thoughtlessly flipping through an entertainment magazine. Dora is sitting across from him on the couch looking at her phone. The door is closed.

MAX: He's a sociopath!
DORA: That's not a thing.
MAX: Sure it is.
DORA: No, it isn't.
MAX: A sociopath is a person who is able to get through the day like everyone else, but who doesn't

know right from wrong.

DORA: That's a psychopath.

MAX: No, a psychopath is a person who can't get even through the day.

DORA: No, that's a psychotic.

Max stares at her blankly.

DORA (cont'd): "Sociopath" is a television word. No one in psychology or psychiatry uses it anymore.

Max continues to stare.

DORA (cont'd): The distinction you're making is between a psychopath and a psychotic.

Max cocks his head.

DORA (cont'd): I minored in Psychology in college.

MAX: Well, he's crazy.

He folds a page back in the magazine and turns it toward Dora.

MAX (cont'd) (off the magazine): Why is George Clooney doing these Nescafé spots? Didn't he marry a barista?

Dora drops her head into her hands.

DORA: A barrister, Max... He married a barrister.

INT. MITCHELL'S OFFICE—DAY

Mitchell is at his desk.

Madison is sitting in a chair across from him. They are staring silently at each other.

INT. DORA'S OFFICE—MORNING

Phil, Madison, Max, and Dora are assembled in Dora's office, two seated on the couch and one in a chair. Dora is behind her desk, giving them a pep talk.

DORA: I know you all know this, but it doesn't seem to be reflected in the work I've been seeing. Now that the whole paper is online-only, Arts & Entertainment—

MAX: —That's us!—

DORA: —will have more room for ourselves, but we'll also be trying to land our stories on the home page just as we did with the paper's front page. So, keep that in mind when you're looking for stories.

MADISON: Don't we have a dedicated spot on the front?

MAX: Yeah, we have a little box on the right-hand sidebar.

DORA: Yes, we have a spot, but I want you to think beyond that.

Max opens his mouth to say something.

DORA (cont'd) (to Max): Don't you dare.

She looks around at the others.

DORA (cont'd): This is a mandate from upstairs. Tell your teams, tell your colleagues, tell your contacts: We want home-page material! [. . .] Any questions?

She looks around again.

DORA (cont'd): Okay, go get 'em! Pitch meeting after lunch.

Everyone gets up, gathers their things, and files out of the office. Max stays behind. As soon as Madison and Phil leave, he closes the door.

MAX: You look beat.

DORA: What have I told you about flirting with me?

MAX: What did you do last night?

DORA: I went out to this show with my son. New local band called Phablet.

MAX: What kind of band was it?

DORA: Info-core?

MAX: Is that some kind of new punk thing?

DORA: Kind of . . . The three members were wearing drum suits and bumping into each other like a mosh pit or something.

Max gives her a blank stare.

DORA (cont'd): According to my son, the patterns they play are based on muscle memory.

Max maintains his blank stare.

DORA (cont'd): It looked more like muscle amnesia. It was like watching Laurie Anderson's illegitimate children slam dance.

MAX: What did it sound like?

DORA: A broken popcorn popper—No, several broken popcorn poppers.

MAX: Wow. What are they called again?

DORA: Phablet.

More blank staring from Max.

DORA (cont'd): It's a portmanteau of phone and—

MAX: —tablet. I know that one.

Dora looks at her computer screen. After a silent moment, she notices him looking at her.

DORA: Hey, I talked to Trace again.

CUT TO:

INT. DORA'S OFFICE—PREVIOUS EVENING

Dora's head is on the desk. She's on a call with Trace. He's on speakerphone.

TRACE (O.S.): I simply cannot believe he still works here, much less that he's supposed to replace you when you retire. [. . .] He's an extra-long phone cord when a cordless is available. He's a dial-up connection when you already have wi-fi. He's a freaking fax machine!

CUT BACK TO:

INT. DORA'S OFFICE—MORNING

MAX: And?

DORA: We're really just going to have to come up with something bigger.

MAX: Bigger than Bob Stephenson!?

DORA: Yeah, I hate the thought of having to run everything through that office before we go live, but I've been fighting with the higher-ups for 30 years. I'm tired. Choosing the right fights is part of surviving here.

MAX: You know, if you'd just retire already, you wouldn't have to fight any of this anymore.

DORA: Oh, and you're ready?!

Blank stare.

DORA (cont'd): Given all of these changes, I don't think you or the rest of the staff are ready.

MAX: Fair enough. [. . .] You're The Man.

DORA: You can still see my breasts, right?

INT. DORA'S OFFICE—DAY

Phil, Madison, Max, and Dora are assembled in Dora's office again after lunch. Phil is pitching a gaming column for the site. He is standing beside Dora's desk with his laptop open to the group.

PHIL: It will complement the other elements of pop culture we're covering in the Entertainment section: TV and movie reviews, celebrity news, etc.

MADISON: And what are you wanting to call this?

PHIL: "Level Up."

MADISON: Hmmm.

PHIL: What?

MADISON: Aren't there several other game-related things named that? Like a podcast or a column in some magazine . . . or a magazine?

PHIL: Oh . . . Maybe.

After a beat, Dora leans in.

DORA: *5150* was the name of Van Halen's 1986 record, their first with Sammy Hagar—

MAX: —after David Lee Roth left, effectively ending the band for all intents and purposes.

DORA: Well, yeah . . . but anyway, in 1992, Eazy-E released an EP also called *5150*.

PHIL Why *5150*?

MAX: It's L.A. Police code for a socio—

DORA: —It's California state code for an involuntary psychiatric hold. Anyway, when a journalist pointed out to Eazy-E that Van Halen had already done a record by that name, do you know what he said?

They collectively shrug and shake their heads.

DORA (cont'd): He said, "SO?!?"

They all look at her sideways.

DORA (cont'd): I used to be a music journalist. [. . .] The name is great. We're not trying to mint a brand here. We're trying to serve our readers. Is there anything else?

They collectively shrug and shake their heads.

DORA (cont'd): All right. See you all this evening. [. . .] Good job, Phil.

After everyone files out of the office, Max points out the window to the conference room at Phil.

MAX: Who was that guy?
DORA: Phil? He's the Web-guy.
MAX: What happened to . . . the other guy?
DORA: His name was Taj, and we lost him to Google over a month ago.
MAX: Ah . . . Collateral damage.
DORA: The damage is only collateral if you end up with what you want.

INT. DORA'S OFFICE—DAY

Madison knocks on the open door as she enters, a notebook and her phone in her other hand.

DORA: C'mon in, Madison. What's up?

Madison looks down at her notes.

MADISON: So far, nothing but rumors on Bob Stephenson.

Max cuts his eyes at Dora.

MADISON (cont'd): I've checked with his management, his representation, his legal team, and his Twitter account.

DORA: Nothing?

MADISON: Well, there are the subreddit conspiracy nuts saying he was killed by the Hollywood underworld or an old college rival, or that he died from injuries sustained after he got hit by a car while riding his bicycle in Reseda, but the only person who responds to me directly is Brendon Walsh.

MAX: His manager?

DORA: If only I could roll my eyes in your language. [...] No. Walsh is the guy who tweeted the thing in the first place, right, Madison?

Madison nods. Max just stares. Dora turns to Madison.

DORA (cont'd): Well, leave me what you have, and I'll see what I can do.

Madison puts her notes on the desk. Max picks them up. Dora throws him a look then looks back at Madison as Madison steps toward the door.

DORA (cont'd): By the way, how did it go with Mitchell?

Dora holds out her hand toward Max.

MADISON: I deleted the photo.

Max hands her the notebook pages.

DORA (to Max): Thank you. (*to Madison*) And?

MADISON: We are in-process.

DORA: What was the photo of?

Madison looks at the floor, then runs out of the office.

DORA (cont'd): Madison!

INT. CONFERENCE ROOM—DAY

As the staff members gather up their things and amble back to their workstations, Dora gestures for Max to hang back.

DORA: I think I solved our Brendon Walsh problem.

MAX: Our who?

Dora glares at him.

MAX (cont'd): Oh, good god, Dora! I know he's the Bob Stephenson guy.

DORA: Keep your voice down, Phablet!

MAX: How? What did you do?

DORA: I hired him.

MAX: You what?!

DORA: Shhhh! I hired him. He starts tomorrow.

MAX: Doing what? Lying on Twitter?

DORA: Kind of... He's going to help with celebrity gossip and social media... So, yeah, actually.
MAX (whispering): What about...

He gestures at the ceiling.

DORA: Trace? Oh, he might have had Stephenson killed, but he didn't say anything about Walsh.
MAX: Speaking of upstairs, have you heard from Tom?
DORA: Not in a few months. No one has.

TAG

INT. DORA'S OFFICE—MORNING

Dora's at her desk drinking coffee. Max comes in just as the phone rings. Dora answers.

DORA: Dora Hondura Lopez. [...] Speaking. [...] Yes, he was supposed to start this morning. [...] I see. [...] Yes. [...] Thank you.

She hangs up the phone just as Madison and Phil enter.

MADISON: Did you hear?
MAX: What?
DORA: I think I just did.
MAX: What?!
DORA: Well, Brendon Walsh won't be joining us today after all.

MADISON: Probably not.

Max looks confused. Dora glares at him.

MAX: Bob Stephenson.

Dora nods. Phil just shakes his head.

MAX (cont'd): High profile. [. . .] Why not?
DORA: He had a bit of an accident.
MAX: What happened?
MADISON: He's in the hospital with a (*checks her notes*) sprained elbow, a broken ankle, and a slight concussion.

Max looks from Madison to Dora.

MAX: What happened?
MADISON: He slipped on a banana peel.

Max's jaw drops.

DORA: I'm never going to get to retire.

FADE OUT.

Fender the Fall

I.

THE HOUSE REVERBERATED IN A LOW HUM. IN the upstairs hall closet, Chris recited aloud, "Up, Down, Down, Left, Right, Left, Right, B, A, Starrr—"

The machine cranked to a start, rumbling... Blinding, bright light engulfed the closet and Chris with it.

In that bright, white flash, the pictures on the walls of the house and the general decor changed. Music now played loudly from a bedroom down the hall. A teenage girl sang along, "as soon as I get my head 'round you, I come around catchin' sparks off you—"

Suddenly, Chris landed on the floor of the closet with a loud thud. The girl screamed, startled. She turned down the music. "What in the—" She peeked out of her doorway into the hall, "Hello?"

Chris scrambled to his feet and stumbled out of the door, naked.

She screamed, he screamed, and the bright light returned, flooding from the doorway. Consumed by it, Chris disappeared again as quickly as he arrived.

The sound of the machine rumbled from the closet. The decor reverted back to normal. The blinding, bright light returned as Chris landed on the floor of the closet with a thud. He scrambled out, clutching his clothes over his naked body. All of the lights were out. He'd blown a fuse.

Chris opened the refrigerator, grabbed a bottle of Lone Star, and headed back upstairs.

Rummaging through boxes and equipment in the hall closet, he found his senior yearbook from 1992: the Lubbock Westerners. He found his picture and chuckled. Then he flipped back to the Sophomores to find his high-school crush, Hannah Reves. He took a pull from the beer. His mind drifted off.

Hannah sitting on a bench outside the Art building with two friends, eating a sandwich. The headband matching her sundress struggles to contain her red curls. The bell rings.

He put the yearbook away and spotted his Nintendo box on the floor under a shelf.

Hannah, alone on the bench, lagging behind the other girls. When she packs up and hurries off, she leaves something behind: a journal with a butterfly on the cover.

Chris shook his head. He grabbed the box and walked out of the room, turning off the light.

Downstairs, Chris noticed a note on the counter.

I've gone to my sister's.
You're never really here, so why should I wait.
I'll get the rest of my things later.

Underneath the note was the journal with a butterfly on the cover. He opened the refrigerator and took out another beer. Then he slid down onto the floor, sobbing softly.

Sitting at the bar at his favorite dive, slowly, sadly drinking a pint of Firemans #4, Chris's grad-school colleague and best friend since high school, Thaddeus, was trying to catch the eye of a woman sitting down the bar, to no avail. Before the thumping funk song finished playing on the jukebox, she was joined by another gentleman.

Thaddeus straightened his faded Bad Flag t-shirt, adjusted the crotch of his skinny jeans, and turned back to his beer. Just then, a blonde woman put her drink down on the bar between him and the door. She pulled a cigarette out of a crumpled pack of yellow American Spirits and stepped outside to smoke it. Thaddeus took a last gulp of his pint and followed her outside.

Outside, the woman held the cigarette between her lips, dialed a number on her phone, and held it between her cheek and her shoulder as she searched her purse for a lighter. Thaddeus lit her cigarette, then his. She nodded her thanks but talked on her phone the whole time they were outside.

He finished his cigarette and followed her back in. Their drinks were gone and all the seats at the bar were occupied.

"They took our seats!" Thaddeus yelled over the music.

"And they took my drink!" the woman added.

"I'll get you another one. What was it?"

She turned and smiled. "Vodka tonic, extra lime."

Thaddeus yelled over her shoulder to the bartender. "Two vodka tonics, extra lime."

"Sarah," the woman said, extending her hand.

"Thaddeus."

They settled into a booth across from the bar, talking over the din as best they could. Thaddeus ordered shots and more drinks. He told her of his Physics graduate studies and found out that she was also a graduate student at Texas Tech, but that she studied Law.

As they continued talking and laughing, Luna, one of Thaddeus's undergraduate colleagues, walked in. Her silhouette in the door revealed spiky black hair atop a wiry frame. She was wearing tight black pants and a shirt cut to reveal several tattoos. She spotted Thaddeus as she ordered a pint. Thaddeus nodded for another Firemans.

Luna put the beers down and slid into the booth opposite Thaddeus. She yelled over the din of the bar, "How're you?"

"If I were any better, you'd be naked."

"Aw, buddy, you know that's never going to happen." She saw the other half-empty glass on the table. "You drinking for two?"

"Nah, she's in the ladies'." Just as he said it, Sarah returned to the booth. "Ah, Sarah—"

"Luna!" Sarah yelled as she hugged her.

"You two know each other?"

Neither answered as they looked at each other then back at Thaddeus.

"Yeah, we—" Sarah started.

"Had a class together," Luna finished.

As more drinks were ordered and downed, the coupling switched and Luna and Sarah were the ones talking, edging Thaddeus out. Many drinks later, the two of them left Thaddeus alone at the bar.

Out behind the Physics building the next morning, Thaddeus and Luna were standing outside smoking, objectifying the female passersby.

"That one?" Luna said, gesturing at a young lady across the quad. "I'd hotbox that one."

"I'd put her out, re-light her, and smoke her all the way to the filter," Thaddeus added as Chris stepped out of the building. Thaddeus spotted another one. "I'd break my own rule and buy a carton of that one!"

"That has to be the worst metaphor I've ever heard by which to objectify women," Chris said, disgusted.

"Speak for yourself!" Luna countered.

Chris turned to Thaddeus. "Wanna get some breakfast?"

"Sure." Stubbing out his cigarette, Thaddeus said under his breath to Luna, "Nonsmoker."

Chris and Thaddeus sat across from each other at their usual spot in the campus cafeteria, breakfasts half eaten. Chris had his books and notes out on the table. On his left wrist, he wore an analog watch. On his right, a digital one. He glanced at them both before pushing his tray aside. "Okay, so what do we know about time travel?" Thaddeus pulled out a flask and took a long pull. "Yikes. What's that?"

"Hair of the dick that fucked me," Thaddeus said. "Is this the Einstein-Rosen portal thing again?"

"No, we're taking that as a given. I'm talking about way, way beyond that."

"Well, that's as far as my knowledge of the subject goes."

"Don't you have a class to teach this afternoon?"

"I'll manage." Thaddeus spun the cap back on the flask.

"Okay... Well, there are three major theories that I think we need to consider," Chris said, referring to his notes.

Thaddeus propped his elbows on the table, his hungover head in his hands. "Correction, there are three theories *you* need to consider. You're making the mistake of assuming that I give a shit about any of this right now."

"First, there's Sparrow's Aberrant Timeline Theory," Chris continued, "which states that a false-tangent timeline can be reconciled by returning to its initial bifurcation. That's all we—"

"You—"

"Need to know. Second, there's the Witness Effect, Nichols and Moon, 1992, which states that an object—a 'witness'—can act as a link between two times or timelines."

"Okay..."

"Like the journal."

"Right." Thaddeus pulled out his flask again and took another deep drink. He spun the top back on. "Okay, what's the third?"

"I knew you were interested! Have you been following the news on the West Texas UFO sightings and alien abduction claims lately?"

"Not really. I mean, I'm aware, but I wouldn't say I've been 'following' them." He gestured air quotes.

"Third relevant theory: Miller's Conjecture, which basically states that flying saucers and time machines are the same thing, a claim all but confirmed by Bosley Crowther in 1960 in *The New York Times*."

"And which are you building?"

"I'm building a temporal projection device," Chris said. Thaddeus just stared at him. "I'm building a time machine. I just don't think mine is the only one anymore." Thaddeus pushed the notebooks away on the table and rested his head on his hands again. Chris continued, "Okay, just think about the inordinate amount of people missing from the reunion this summer, the return of the Lubbock Lights and the terrorist attack last fall... There's something going on! Something has gone terribly wrong!"

"No duh, Yoda! I just don't see how all of this is relevant to me, you, or anything related to us!"

"The thin veil between us and parallel universes or timelines is thinner now than it usually is, easier to penetrate. It's what the Celts would call a 'Thin Space.' Something has weakened the structure of spacetime here or near here. The global weirdness and so-called 'abductions' are the clearest sign yet!" After a moment, Chris leaned back in his chair and smiled.

"What?" Thaddeus said.

"It works." Chris thought for a moment, then added, "Well, I think it does."

Thaddeus nursed his hot coffee. "What, your Nintendo?"

"No, my machine."

"Your time machine? Get out of here!"

"Shhhh! No, really. I think it works."

"What do you mean 'you think'? What happened?"

"There are still a lot of kinks, but I went to my house in 1991 and I was naked and . . ."

"Are you sure you didn't just bump your head real hard?"

"Yes! I went back to 1991, and I ended up in someone else's house, which is my house now."

"If that didn't make flawless sense, I'd tell you how crazy I think you are right now." Thaddeus stared at Chris for a moment. "What were you trying to do?"

"I was trying to go back to high school and give Hannah her journal back."

"Okay, now I *do* think you're crazy! When I said get rid of it, I meant, like, throw it away or something, not travel back in time and return it to its rightful owner!"

"Yeah, but if I give it back to her in the past, then I never had it!"

"Again, with the flawless logic—"

"If I never had it, then my wife never leaves."

"What?"

"I've run all the scenarios. If Hannah doesn't forget her journal that day, then my life doesn't fall apart yesterday."

Thaddeus gave him a skeptical look.

"The problem is, according to my initial test run yesterday, I can't take anything back with me: no physical objects, no food, no clothes, no jewelry, nothing. So, I can't take the journal. Only my body can go—"

"The journal is already there."

"What? Oh, man . . . You're right! I just have to make sure she doesn't forget it and that my younger self doesn't get it!"

"Exactly," Thaddeus agreed.

II.

Chris's clock radio was between stations and blaring static. It was 7:21 AM. He reached over to shut it off and noticed anew that he was alone in the bed.

In the hall closet, over the low hum of the machine, Chris dictated into a mini-cassette recorder: "Given the experimental nature of this project and the inability to transfer lab notes and other materials as discovered in the initial test run, I am recording these sessions from now on." He adjusted the controls and set the recorder on a shelf among the machine's many components. "I am Christopher Bridges, Ph.D. candidate in Theoretical Physics at Texas Tech University." He took a drink from a half-empty cup of coffee. "I have assembled a device for the purpose of time travel via temporal projection. I have successfully visited the past on one occasion so far. This will be the second test." He put the cup down beside the recorder and turned a dial on the machine.

"I am setting the chronometer for 11 AM, Sunday, September 1, 1991." He turned two dials and flipped a switch. The machine hummed louder. He picked up the Nintendo controller. "Initiating projection sequence: Up, Down, Down, Left, Right, Left, Right, B, A, Start... Up, Down, Down, Left, Right, Left, Right, B, A, Start—"

The machine popped and sputtered, and all the lights went out.

Chris hung his head and dropped his notes and the recorder on the floor. "Dammit."

III.

Chris was finishing up a lecture in his Intro to Physics class. His students were slowly trying to quietly pack up their books.

Chris turned from writing on the chalkboard. "All right." The students paused their packing. "Review the first two chapters in your book, as I know you've already read them, and read chapter 3 for Thursday. See you then."

The students scrambled out of their seats. Thaddeus stepped in just as the students crowded the door. Chris was packing up his notes. He didn't look up as Thaddeus approached him at the podium.

"How may I help you?"

"What was the homework, again?" Thaddeus asked in a meek voice.

Chris looked up and frowned.

"Yikes. What's up, man?"

"I don't want to talk about it."

"C'mon. I'll buy you lunch."

Chris picked up his bag. "I'm not hungry."

"You gotta eat, man."

In the cafeteria, Chris pulled out a crumpled piece of paper with notes scribbled all over it. "It stopped working."

"What, your digestive system?"

"No. The machine."

"That's probably for the best." Thaddeus gestured at the notes with his fork. "Are these your notes?"

"Yeah, something like that," Chris flattened the paper on the table. "Here's what I've learned from my research and the one brief trip I was able to take." He counted points off on his fingers: "1. You can't take anything back with you, including clothes. If it's not a part of your corporeal body, it stays in your native present. 2. You're not really traveling through time but being projected to another point in time. That part is confusing in that it doesn't follow the way we normally conceive of time travel and that it leads us to, 3. Hours spent in one time are taken from the other. So, if you're in another time for 24 hours, you're not anywhere else for those same 24 hours."

"I wondered about that," Thaddeus said. "So, you can't just come back to the moment when you left?"

"No. It's like the temporal entropy in McKenna's Timewave. There's conservation of time the same way there's conservation of energy. Memory is heat, and heat is the source of time—"

"How long do you need to be there?"

"Not long, but I'll only have the one chance. None of it matters if it doesn't work on the day." He looked out the window. "I'm going to have to move out, man. The house is hers."

"I know." They ate in silence for a while. Thaddeus noticed something among Chris's notes and books. "What's this?"

"A calendar from 1991."

"When exactly are you trying to return to?"

"Well, September 9, 1991 is the day Hannah leaves—"

"Left—" Thaddeus corrected.

"Her journal on the bench outside the Art building."

Thaddeus pulled out the calendar and took a closer look. "When did you go back to last time?"

"I didn't set the day. I didn't think it was going to work, and I got sent back before I was ready."

"When was that?" Thaddeus asked.

"Two days ago?"

"So, Sunday, September first."

"Yeah? So?"

"I don't know, man. I'm just trying to help." As Thaddeus finished eating, Chris stared at the calendar. He had his elbows on the table and his head in his hands. He hadn't touched his food. Thaddeus swallowed his last bite. "Did you set the time for the trip last time?"

"What do you mean?"

"I mean, you said you didn't set the day. What did you put into the machine for your destination?"

"I didn't. The only thing I adjusted for was the year, 1991."

"Okay. Well, I obviously don't know enough about it, but it seems like that might be part of the problem."

"Which part?"

"The settings."

"I appreciate your help," Chris added.

IV.

The machine's low hum resonated through the top floor of the house. Chris talked over it into his tape recorder. "—unlike on my initial test run, I am setting the destination to the minute. I am setting it for the exact same time as the last projection: 11 AM, Sunday, September 1, 1991." He put the recorder down and picked up the controller. "Initiating sequence... Up, Down, Down, Left, Right, Left, Right, B, A, Start... Up, Down, Down, Left, Right, Left, Right, B, A, Start—"

The machine cranked up, popped, and sputtered, and all the lights went out again. Chris hung his head and sobbed silently.

V.

Chris, looking slightly more disheveled than usual, came out of class behind his students and turned into Thaddeus's office. "Hey. You were wrong."

Thaddeus turned from putting papers in a filing cabinet. "About Dr. Bailey and his T.A.? Nah, man, I'm pretty sure about that."

"No."

"Then you're going to have to narrow it down. I'm wrong about a lot of things."

"The time settings on the machine."

"Right." Thaddeus turned from the files.

"It still doesn't work."

"I think there's still something there. Early lunch?"

Chris was staring at the September page of the 2002 bikini calendar hanging on Thaddeus's office wall.

"Hello? Early lunch?"

Chris snapped out of it. "What? No. I need to get back to this." He turned to go.

"Whatever, man. See you sooner."

Chris stopped in Thaddeus's doorway. He shuffled through notes and papers in his arms until he found the 1991 calendar. He flipped it to September. He looked at the page, astounded, then turned around and pushed past Thaddeus. He held the 1991 calendar up next to the 2002 one, both showing the month of September. "You weren't wrong, but you didn't know you were right."

"That doesn't sound like me," Thaddeus said, looking at his watch. "About what?"

"Look! They're the same! Identical calendar years!"

"Freaky!"

"Yeah! It's a convenient coincidence I hadn't counted on. Fortunately, I—"

"We—"

"Figured it out before the crucial day passed."

"Hey, aren't you defending your prospectus soon?" Thaddeus asked.

"Yeah, on Monday."

"And don't you have enough data now to finish your dissertation?"

"Probably, but you know that's not the point. I want my life back." Then he looked at Thaddeus, "I want my wife back."

VI.

As Chris pulled into his driveway and turned off the engine, the needle on the gas gauge settled firmly over the E. Chris got out and grabbed his notes and books from the backseat.

Hours later, books and notebooks piled on the kitchen counter and dining room table, Chris sat drinking a beer, and double-checking his notes. The parallel calendar days were open on top. Then Chris pulled out his notes on the power requirements of the machine.

He headed out to the garage and dug a gas-powered electric generator out of the boxes in the dim light. He checked the fuel: half a tank. "That will have to do," he said as he replaced the gas cap. He grabbed the generator and hauled it out the back door.

He fell asleep at the dining room table, his head on a pillow of open notebooks, various books, and papers.

VII.

Chris woke up, slowly realizing where he was. He saw in his notes that he'd written something about the power outage saving him last time.

He made some coffee and took a shower.

In the hall closet, Chris talked to his tape recorder. "Seemingly due to the conservation of time on which the mechanism is based, the machine only returns to the same time in parallel years. These happen in odd intervals of 6, 11, 12, 28, and 40 years. This year is in between two 11-year cycles: 1991, 2002, and 2013." He set the recorder and his notes down again. "I am setting the year for 1991. If everything works properly this time, I should arrive there at exactly the time it is now but in 1991: seven days after the last trip: 11 AM, Sunday, September 8, 1991, and return tomorrow evening: 5 PM, Monday, September 9, 2002."

Chris looked at his digital watch. It read 10:59 AM. "Initiating sequence. Up, Down, Down, Left, Right, Left, Right, B, A, Start... Up, Down, Down, Left, Right, Left, Right, B, A, Start..."

The machine cranked to a start, humming, rumbling... a blinding, bright light consumed the room. It faded as fast as it flashed, leaving Chris's two watches and his wedding band on top of his clothes piled on the floor of the closet.

The decor in the hall returned to its past luster. Music played loudly from the bedroom at the end of the hall. A teenage girl sang along.

Suddenly, blinding, bright light shone out of the doorway as Chris landed on the floor of the hall closet with a thud. The girl screamed, startled. She turned down the music and slowly emerged from the room with a baseball bat. "Hello?"

Chris scrambled to cover himself with pillows. "Hello."

"The hell? Oh, wow... I thought I'd imagined you!" the girl said.

"You didn't," Chris said, awkwardly standing in the closet doorway.

"I have the most wicked sense of déjà vu right now," she added.

"That means you're on the right track. Déjà vu is a sign that your decisions have aligned you with the one true path."

She just looked at him. Then suddenly brandished the bat. "Wait! What is this? How did you get in here? Again?"

"Please, stay calm. It's a long story," He said, clutching the pillows. "Where are your parents?"

"Church." she started unsure then turned threatening. "But they'll be back any minute!"

"Oh yeah, it's Sunday."

"Yeah."

"What time is it?" Chris asked.

"Time for you to make like a tree and get out of here!"

Chris just stared at her.

"No, really! Time for you to go."

"No, really. What time is it?" he insisted.

"I don't know! It's about 11." She loosened her grip on the bat. She was smirking at Chris and his strategically-grasped pillows. "Let's go, mister."

"Do you mind if I cover up a little better before you run me out of here?"

"I don't care if you go clothed or naked, as long as you go!"

"Do you really want Ms. Myers to see a naked man creeping out your back door while your parents are at church?" he said, nodding toward the neighbor's house.

"Fair point," she said. "Wait. How do you know Ms. Myers?

"Get me some clothes, and I'll tell you."

In the kitchen, Chris and the girl were having tentative coffee. Chris was dressed in her father's ill-fitting clothes. She was still clutching the baseball bat. Pressed for time, he was attempting to explain the situation to her before she ran him off, beat him to death, called the police, or her parents got home.

"You came all of this way just to return her silly journal?" she asked.

"Yeah," he said as he reached for the sugar. She flinched slightly at his movement, then pushed the sugar across the kitchen island toward him with the end of the bat. "Wouldn't you want yours back?"

"Well, yeah, but it just seems like a lot of work for one girl's journal." A moment passed during which they sipped their coffee in silence. She relaxed some, but still regarded Chris with hesitance.

"I mean, there's a lot more to it than that, but let's just say it's about peace of mind."

"Whose? It makes you seem crazier than just someone trying to get some sleep." she sipped her coffee. "Is this like a *Back to the Future* situation?"

"Not exactly—"

"I loved that movie in middle school!"

"Well, it's kind of like that." He adjusted her father's shirt. "I can't tell you everything that's at stake, but it all goes down tomorrow. Will you help me?"

"Maybe," she said, mocking reluctance, "but you have to tell me the whole story eventually."

"Deal."

She thought about it for a minute and then added, "I'm Samantha, by the way."

"I know," he said. "You're Samantha Sever."

At the Bridges' house in 2002, the phone rang twice. The answering machine clicked on. Chris's voice said, "You've reached the Bridges, please leave a message."

A woman's voice started after the beep, "Chris, I know you're upstairs working on your project, but I do wish you'd answer the phone. We need to talk about this. Please, call me back."

Back in 1991, Chris and Samantha continued their tentative coffee conversation.

"So, what exactly is your plan?" she asked.

"Well, I need to get to that bench just outside the side entrance to the Art building when Hannah forgets her journal—"

"And make sure she doesn't."

"Right. If she never forgets the journal tomorrow, then Young Chris never picks it up—"

"And then Old Chris—" she said, pointing at Chris.

"'Chris' is fine—"

"You, never has it."

"Exactly."

A silent moment passed between them.

"It's crazy, but it's imminently doable," she smiled deviously. "If I were to help you . . ."

"Your help would be invaluable! Essential!"

"If I were to help you . . ."

"Okay. If you were to help me, what would you require?"

"I want to go to the future."

Another silent, tense moment passed, then Chris turned very serious. "I don't think that's possible, Samantha. The machine is there, in 2002, you are here in 1991." Samantha crossed her arms, suddenly pouty. "It's also very dangerous." Seeing that she was unconvinced, he added, "I will look into it when I get back, but I don't want to get your hopes up."

"You're going to go back without me, and I'm never going to see you again!"

"I promise I will see what I can do, Samantha. Will you just please help me get Hannah's journal back to her?"

"Maybe . . ."

"Please, Samantha. I will reciprocate in an appropriate manner." Samantha side-eyed him with contempt. "Starting with seeing if I can bring you to the future. Okay?"

Another moment passed, a little easier this time, during which Samantha hesitantly uncrossed her arms.

"Okay."

VIII.

That evening, Samantha sat at the dining room table doing homework. Chris sat across from her thinking about the plan. He scribbled notes on a piece of her notebook paper. Several minutes passed in silence.

"'Antediluvian' and 'prelapsarian,'" Samantha said, "Which is 'before the flood' and which is 'before the fall'?"

Chris looked up at her. "I don't know."

"Seems like something you should know," she said, returning to her books.

They sat in silence for another moment. Then Chris turned to Samantha, nodding at her school papers. "What is this?"

"Vocabulary. What is it with you?"

"Nothing," he said, looking away again. "What do you usually do for fun?"

"I don't know what makes you think there's anything fun to do in this town on a Sunday night."

"Sure, there is!"

"How would you know?"

"I grew up here, too, remember?" he said proudly. "Once a Westerner, always a Westerner."

"Duh!"

"No duh, Yoda." They both laughed. "My friend Thaddeus says that all the time."

"Thaddeus Noble?" Samantha asked.

"Yeah. Wait, you know Thaddeus?"

"*You* know Thaddeus?"

"Yeah! He's my best friend," Chris said.

"Weird!"

They sat in that weird silence for a moment.

"Let's get out of here," Chris said. "Do you have bicycles?"

"Yeah, but I'll have to be back before my parents get home."

"Are they ever coming home?"

"Yeah, in a couple of hours. I lied before to scare you."

"No problem. I need to find a place to stay the night anyway."

IX.

Thaddeus and Luna crept around to the back of Chris's house. Thaddeus was about to check under a potted plant for the spare key, and stopped dead, noticing something in the foreground: a ring of dead grass, in a perfect circle, in the middle of the otherwise overgrown backyard. "What in the—" They stared for a moment, stunned, before returning to find the spare key to unlock the back door.

Thaddeus looked over his shoulder as he unlocked the door. "You don't think…"

"Think what? That he was abducted in a flying saucer?"

Thaddeus shook his head. "Never mind."

As Thaddeus and Luna entered the back door, they immediately called out for him.

"Yo, Chris!"

"Mr. Bridges?"

They looked around the room. It was a wreck. There were books and notes strewn on every available surface, accompanied by empty pizza boxes, soda cans, and beer bottles.

They stood, baffled. Thaddeus said, "Let's find his lecture notes and get out of here. It looks like a train wreck and a crime scene had sex in here."

X.

Chris stood in the driveway between two bicycles as Samantha ran back into the darkened garage to push the button on the garage door opener. The door lurched and grinded, slowly closing as she ran and ducked under it just in time. She grabbed the handlebars of one of the bikes.

"Let's go."

They pedaled off down the road, down a light hill, the road only visible in the circles illuminated by the sparse streetlights. She weaved, clowning, laughing, cutting him off, and riding with no hands. At the bottom of the hill, she took a sweeping left, nearly losing him.

Samantha pulled ahead of Chris over the next incline.

Chris rode the brakes, screeching as they came down the steep hill. Samantha leaned into the wind going faster and faster. Halfway down, Chris trailed her by a few car lengths.

Then he noticed the stop sign at the bottom.

Then he noticed a car coming in fast from the left.

He yelled, trying to get Samantha's attention, but she was tucked in, zooming toward the intersection.

The car bore down from the left.

Samantha ran the stop sign, cackling wildly and cleared the road just before the car blew through.

Chris zoomed through just after the car passed, his brakes still screaming.

The car stopped and backed up, but they were already gone in the darkness.

Chris caught Samantha at the top of the next hill and gave her a stern look. She just laughed and pedaled off down a trail. He followed off the road through some light brush and trees to the top of another hill. She dropped her bike and climbed up on top of a big rock.

"You almost got me—" he said, panting, "us killed back there!"

She ignored him.

"I love it up here," she said, sitting down.

Chris laid down the other bike and stood beside the rock. They both looked up at the stars. "Isn't this the place?"

"What place?"

"The place where the Lubbock Lights were first spotted?" Samantha just looked at him. "In August and September of 1951, several people, including four professors from Texas Tech, saw a V-shaped formation of lights fly over here."

"Oh yeah! That happened again last year! They said it was just birds, though."

"Yeah, migrating plovers, flying in formation."

"Yeah!" she said.

"Wait, you said it happened again last year?"

"Yeah, why?"

"Hold on." Chris looked at the sky, counting on his fingers, doing math in his head. "There must be a six-year jump in there somewhere."

"What?"

"I was trying to figure out if 1990 and 1951 are parallel years—years with identical calendar days." Samantha tilted her head at him, thinking hard. "Like 1991 and 2002. The reason I was able—"

"No, I got that part. I was trying to find the six-year gap. We need actual calendars, though."

"If 1990 is parallel with 1951, then 2001 probably is too."

"Yeah, and?"

"I can't tell you."

"Dammit. This sucks. You have to give me something!" Chris looked down at the ground. Samantha looked back at the sky.

"Okay. Something really bad and bizarre happens in September of 2001, but that's all I can really tell you."

"C'mon, Mister!"

"Just stay out of New York and Washington, DC during September of 2001."

She looked mildly confused. "Okay." They looked at the sky in silence for a moment.

"I do wish we could figure out the calendar correlation between 1951 and our adjacent years."

Samantha jumped down off the rock. "Me too." She picked up her bike and hopped on, coasting down the hill. Chris snapped out of his reverie and followed.

They rolled up to the playground behind the elementary school next to the high school. Samantha ditched her bike and ran for the swings. Chris followed, sitting on the swing two over from hers.

"Did you go to Lollapalooza?" he asked as he turned the swing toward her. She started swinging.

"No. I couldn't get a ride to Dallas for either show!" She swung past a couple of times. "Did you?"

"Yeah, 1991 and 1992, but none after that."

"Oooh! Who's on it next year?"

"I guess I can tell you that. Let's see... The Jesus and Mary Chain—"

"Awesome! I love them!"

"Soundgarden, Pearl Jam... Ice Cube... Ministry... Red Hot Chili Peppers headline."

"Less awesome," She swung past again. "Who's on it in your year?"

"2002?"

"Yeah."

"Oh, they stop doing it in '97."

She swung in silence for a moment. "Weird."

"Yeah, it gets really weird before it finally just stops."

On her next pass, she noticed his left hand gripping the chain of the swing. His finger marked by a missing wedding band.

"So, you're married?" She kept swinging.

"Yeah, for almost seven years now."

"How did you two meet?"

He thought about it. "I can't remember the moment we met. I had a crush on her for a long time though."

"When did you fall in love with her?"

"That I remember. We were at a party. I was half-drunk and trying to bum a cigarette from our mutual friend Katie, when she approached with the same plan. When Katie offered each of us one, she said, 'No, Chris and I can share one.' That was it. I was done for." He watched her swing by again. "Neither of us has smoked since, but we've split everything else."

"When did she fall in love with you?"

"I have no idea."

She swung by in silence again. "Wow. What's that like? Being with someone for so long?"

"Hard. Fun. Easy... Confusing."

"It sounds exhausting," she said, kicking the ground.

"It is. It's forever, but you always question whether or not it's really forever. Other people get in the way."

"What do you mean?"

"Other people. Interests. Crushes. Et cetera."

"Ooooh, Mister. Crushes?" she giggled.

"Yeah."

"Your heart's so big, you just want to love all the girls?" she mocked.

"Very funny. That's part of why I'm here. Now."

"That's why I have no desire to get married." She looked at him as she passed. "I'm going to have questionable relationships based mostly on sex until I'm old enough not to care anymore, then I'm going to marry someone else who feels the same way."

"You are wise beyond your years, Samantha Sever."

She shifted her voice as if quoting someone. "'I am a grown woman with a lifetime of experience that you can't understand.'"

"What's that?"

"*Peggy Sue Got Married*. Another favorite circa middle school." She dragged her feet, stopping the swing. "Oh, shit! What time is it?"

He looked at both of his wrists. "I don't know."

"I need to get home." She jumped off the swing and headed for the bikes.

"What about the other bike?"

"I'll get it tomorrow." She swung her leg over the top tube and pedaled away. "See you then!"

Chris suddenly panicked. "Samantha, wait!" She carved back around to the swings. "Do you have any ideas about where I could spend the night?"

"Oh yeah, you don't have any money, huh?"

"Everything I have, you have given me," he said.

She thought for a moment. "I could probably hide you in the garage like E.T."

"You think?"

"It might not be the most comfortable place, but it won't cost you anything but a trip to the future." She smiled.

Her parents' car was already parked in the driveway when they got there. Samantha whisper-shouted, "Shit! They're already back!" Chris ducked. "Here. Wait around the back by the door to the garage. I'll come get you when it's clear. It shouldn't be long."

Chris pushed the bike around the side of the garage as Samantha snuck inside.

A few minutes later, Samantha popped her head out of the back door to find Chris sitting against the wall next to the bike. "Psst!" she whispered. Chris jumped up. "I changed my mind. Here's some money. There's a hotel on—"

"34th—"

"Yeah, on 34th. This should be enough for tonight."

Chris held out the wad of bills. "Where did you get this?"

"I 'borrowed' it from my mom's purse."

"Yikes. Thank you!"

"Shhh! I'll get the bike tomorrow like I said. Now get out of here!"

Unable to sleep, Chris tossed in his hotel bed. He found the remote control and turned on the television. He flipped past a special report on Anita Hill's sexual harassment case against Clarence Thomas and stopped on a news story about child abductions. "Everywhere I go, people are disappearing," he said to the flickering screen.

Samantha woke up to loud music. She got up and went straight to the shower.

Chris woke with a start after a fitful night of little sleep. He slowly moved toward the shower.

Samantha dug through her clothes, slowly getting dressed, singing along to the loud goth-pop. She found her VHS tape of *Back to the Future* and smiled. She put it on her dresser next to her VCR and TV. *Peggy Sue Got Married*, *Beetlejuice*, and *E.T.* were also stacked there.

Chris put on Samantha's father's ill-fitting pants, button-up shirt, and too-big shoes.

Samantha ate cereal and looked back over her vocabulary homework.

Chris gathered up his notes from the night before and headed out of the hotel room.

Samantha put her cereal bowl in the sink, put her notes in her backpack, slipped it on, and headed out the side door of the kitchen.

Nervous and stressed out, Chris waited at the bus stop. An old woman approached, carrying an Octavia Butler paperback.

"Excuse me, son, do you know what this word means?" She was pointing to the word "latent."

"It means inactive, or in a holding pattern."

"Oh . . . Thank you." She walked away, satisfied.

"Yes, ma'am."

Young Chris climbed onto the crowded school bus. He was wearing a backpack and a Walkman, its spongy orange headphones mushed against his ears. The first empty seat was beside Young Thaddeus. Chris slid in just as the bus lurched from the curb.

"What are you listening to?" Young Thaddeus said, not looking at Young Chris.

Young Chris pulled off his headphones.

"What are you listening to?"

"*De La Soul Is Dead*."

"Ring Ring Ring Ha Ha Hey!" said Young Thaddeus. "Yeah! So good!"

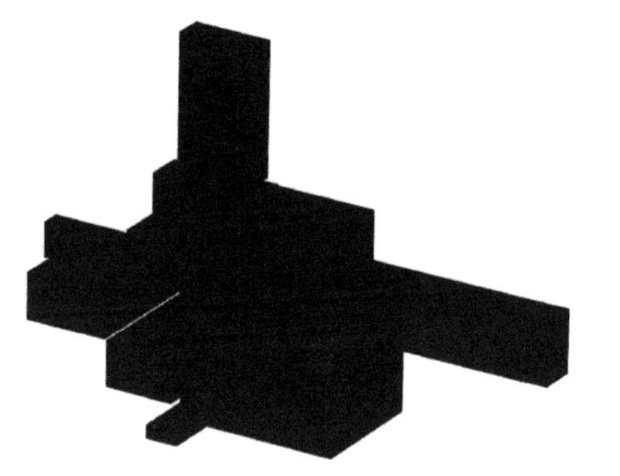

XI.

In Chris's Intro to Physics class in 2002, a classroom full of students sat restlessly at their desks, some chatting quietly, others checking their watches impatiently.

Finally, they gave up in groups and filed out of the room.

Thaddeus and Luna were silently eating breakfast together.

Chris's office was empty and silent as students gathered outside the door, milling about in the hallway. One knocked on the door to no avail.

XII.

The bus dropped Chris off across the street from Lubbock High School. He jogged across the street behind the bus as it left and walked briskly around the corner in view of the Art building.

He thought hard. It was 3rd period. He had Art class 3rd period with Thaddeus. Across the courtyard he saw the bench outside where Hannah ate her lunch.

The bell rang, and students spilled into the courtyards between buildings. In the chaos. Chris didn't see Principal Carter approach him from behind.

"Excuse me. Can I help you?"

Startled, Chris stammered, "No, sir. I'm just here to meet someone."

"Oh? Well, all visitors must register with the front office."

"But I'm just here to return something."

"Be that as it may, all visitors must register with the front office. Come with me, sir."

"But..."

As Principal Carter walked into the building, Chris made a break for it, his too-big, borrowed shoes flopping like a clown's. As he rounded the corner of the building, he glanced over his shoulder. He saw someone else peeking around the corner. A haggard older man clothed like the homeless. Panicked, startled, and confused, Chris kept running.

Samantha waited in her designated spot. She had skipped lunch to help Chris, and he never showed up. Now she was going to be late for 4th period. She stomped off to class.

On her way to Ms. Morgan's 4th period English class, Samantha spotted Young Chris Bridges at his locker. She made note of the number.

In class, Ms. Morgan was lecturing. Samantha looked out the window.

"Dana is worried about Kevin being affected by the 19th century," Ms. Morgan explained, "and how do we see that he is?"

"He seems quieter, angrier," a student answered.

"Maybe more aware of the realities of the racialized past?" Ms. Morgan prodded. "The novel also begins with a striking example of foreshadowing. Can anyone tell me what it is?"

The discussion was interrupted by the crackling of the public address system: "Samantha Sever, please come to the office. You have a telephone call."

To the students around her, Samantha quipped, "Must be my agent." She got up to go as Ms. Morgan pointed to the door, visibly irritated.

Chris stood at a payphone a few blocks from the high school. Dropping loose change from his pocket, he picked up the only quarter he had.

Samantha was already annoyed with him. "Hello."

"Samantha! Oh, thank god! You gotta help me."

"Hi, Mister. Where have you been?"

"I got caught outside the school by Principal Carter!"

"Well, I waited for you at the meeting spot, and when you didn't show, I had to run to Ms. Morgan's class. I was late again, and my excuse would've sounded ridiculous, so I didn't give her one."

"Did you see Hannah? Did you get the journal?"

"No! I told you, I left! I didn't even see her! I was waiting for you!"

"Where was I?"

"You tell me!"

"No, younger me!"

"I didn't see you, er, him! I left!"

"You have to go back and try again!"

"No, I don't." Samantha hung up the phone.

"Samanth—" Chris dropped the phone. He couldn't stand still. He paced the sidewalk, stomped, kicked, pulled at his hair, looked out at the cars going by. He stopped suddenly in the middle of the sidewalk.

After a moment of complete stillness, the sidewalk scene started to ripple, and Chris disappeared from the sidewalk in a bright, white flash of light.

XIII.

The high school hallway was eerily empty and silent.

The bell rang, and students flooded the hall with bodies and noise.

Leaving Ms. Morgan's English class, Samantha briefly staked out Young Chris's locker, #127. She pretended to try to open the one next to his. Young Chris approached but didn't seem to pay her any attention as he opened his.

"Forgot my combination," she said in his direction.

"That's not your locker," he said without looking at her.

"You're right," she improvised, "It's my friend's. I was supposed to get a book out of it. Do you have Ms. Morgan for English?"

"No. I have Dr. Strength."

"He sounds like some sort of superhero or something."

"Yeah. Too bad he's not." He closed his locker. "I have to go."

"Wait."

Young Chris stopped and turned back toward Samantha.

"Never mind," she said as he started to walk away. "You always look like you're looking for someone."

He turned around, briefly walking backwards. "Maybe I am."

XIV.

Chris landed naked on the floor of the closet on top of his clothes, watches, and ring from the day before. All the lights in the house were off again. The generator ran out of gas, the fuse blew, and the machine shut down, sucking him back to the present. He'd been gone almost a full 24 hours.

He scrambled to his feet, clutching his clothes. A terrified look crossed his face as he saw it: Hannah's journal was still there.

He started to get dressed, and then his face folded into another terrified look as he heard floorboards creak. Someone else was in the house.

"Hello?"

He waited. When there was no answer, he finished getting dressed. After searching the house, he headed outside.

Chris checked the generator. It had indeed run out of gas. He got an empty red gas can and a hose from the garage and pushed the button to open the garage door. It didn't open. "Oh, right. No power." He walked out the back door and around the house to the driveway. He struggled to siphon the remaining gas from his car's tank. After several attempts, he finally got a slow trickle to flow into the gas can for a few seconds.

"That will have to do."

XV.

Samantha went to the locker next to Young Chris's again after her next class, pretended to know the combination, and tried to open it. She said the numbers aloud as she turned the dial. "28 . . . 6 . . . 42 . . . 12 . . ." She pulled on the handle, feigning surprise when it didn't open. "Hmmm . . ." She turned the dial again. "28 . . . 6 . . . 42 . . . 12 . . ." Again, the locker didn't open. She gave up and turned to lean against it. Young Chris still hadn't shown up. She waited a few more minutes. There was no sign of him.

The bell rang.

XVI.

"She has a minor in Passive Aggression," Luna said, taking a long drag off of her Parliament.

"Well, you can't major in it," Thaddeus added, lighting up.

It was mid-morning on the Texas Tech campus. An attractive female student walked by on the quad.

"I'd slam dunk that one," Thaddeus said, exhaling smoke over his head.

"I'd do the grand slam on her!" Luna said. Thaddeus threw her a sideways look. "What? A grand slam is better than a slam dunk."

"You want to be the fourth one in? It's also not even the same sport!"

"I guess I should leave the sports metaphors to you."

"Please."

XVII.

Chris replaced the fuse and fueled up the generator with what little gas he could get out of his car. He was in the hall closet again. He saw the recorder on the floor and kicked it away.

"No time for keeping records now."

Just as he punched in the launch code, he heard the floor creak and the thump of someone else in the house again. It was too late. He disappeared in a bright flash.

A woman emerged from his bedroom wearing his clothes. "Godspeed, Chris Bridges." She walked into the closet, and flipped a few switches on the machine, adjusting cables and cords as she did. "He really needs to update this thing. This version is not safe."

She set the date for 2019, and punched in the launch code, then came the bright flash, and she was gone.

XVIII.

The decor in the hall again reverted to its former style and arrangement.

After the bright flash, Chris landed on the floor of the closet. "Like the flapping of a black wing," he said to himself. This time when he stuck his head out into the hall, no one was home. "Okay. Monday afternoon."

The bus dropped Chris off across the street from the school again. Wearing another ensemble of ill-fitting business clothes, he jogged across the street behind the bus as it left and walked briskly around the corner in view of the Art building.

It was 5th period. The students were all in class. Across the courtyard he saw the empty bench.

The bell rang, and Chris didn't see Principal Carter approaching him from behind again. "You again." Chris turned around, just as startled as he was the first time. "We've been over this, sir—"

"I know. I'm just here to return something."

"There have just been too many abductions, missing children, and weird visitors lately. We can't afford to look soft on this." Principal Carter grabbed Chris's arm. "I don't know how you got here again, but you're not welcome."

Chris snatched his arm free and started to run again. Principal Carter managed to grab him this time.

Samantha staked out Young Chris's locker after 5th period. She pretended to open the one next to his again when Young Chris arrived. "28 . . . 6 . . . 42 . . . 12 . . ." It didn't open.

Trying to ignore her, Young Chris set to work on his own combination. He opened his locker and began swapping out books for his next class. Samantha spotted Hannah's journal in his locker.

"What's that?" she asked.

"What's what?"

"That," she said, pointing at the journal.

"Nothing."

"It's not nothing, and it's not yours." Samantha grabbed the journal just before Young Chris slammed the locker shut. With wide eyes, she looked at the cover. Young Chris tried to grab it. Samantha quickly pulled it away and ran off down the hall.

"Hey! Come back here!" Young Chris cut down the adjacent short hall and down the second to head Samantha off on the last one. He stepped out in front of her. "You've got to give that back."

"It's not yours," Samantha said as she tried to step around him.

Young Chris stepped in front of her again. "It's not yours either."

Samantha stepped back the other way. "You need to return it to Hannah."

Young Chris stepped in front of her again. "I can't."

"I can." Samantha pushed past him and disappeared into the students crowding the hallway.

Principal Carter was sitting at his desk with Chris seated across from him. He didn't believe Chris's made-up story about meeting Samantha. "I'm afraid we're going to have to bring in the authorities. Like I said, we can't afford to look soft on this."

"Look at me!" Chris said, standing up. Principal Carter looked him over, his oversized shirt, pants, and his too-big shoes, and made a doubtful face. "Do I look dangerous?"

"We'll let the police decide that."

Chris made a break for the door. He ran past the receptionist, out of the office, and toward the front door of the high school.

In the front hall, he ran past a woman dressed like a teenager in a white button-up and a polka dot skirt, on her way into the school. She could pass for a teacher, but he knew she wasn't. The polka dot skirt she was wearing seemed familiar to him, and it threw him off. She smiled and held the door open for him. He looked back over his shoulder, brow crumpled in confusion.

He ran out of the front door just as the police pulled into the parking lot. He turned back toward the school, fumbling backward. The woman was still smiling. He was quickly taken down and into custody by the police.

The woman walked in the front door and confidently entered the front office. She approached the receptionist with purpose.

"Hello, ma'am. How may I help you?" the receptionist asked.

"Hi. I need to see Thaddeus Noble."

"The receptionist handed her a clipboard and set a cup of pens up on the counter. "What is this regarding?"

The woman put the clipboard down on the counter without looking at it. "I really just need to give him a personal message. It's about his ride home today."

"Okay, ma'am," the receptionist said hesitantly. "Give me just one moment."

She knocked then went into Principal Carter's office and pulled up the door behind her. The woman could hear the murmurs of their conversation through the door. When she came back out, she left the door open. Still visibly agitated from his run-in with Chris, Mr. Carter nodded at the woman. "One more moment please." The receptionist scribbled on a notepad and sent a student office worker to get Thaddeus.

A few moments later she returned with Young Thaddeus, a hesitant nerd, missing the hipster confidence he grew into later. The woman stepped out into the hall to meet him. The receptionist and Principal Carter watched them talking through the window in the office.

The nonverbals of their conversation escalated quickly, both of them gesticulating aggressively. The whole scene slowed to a stop. Just as Principal Carter approached the door, the hallway rippled, and the woman disappeared in a bright, white flash of light. Principal Carter walked out to console an obviously shaken Young Thaddeus.

Chris was sitting next to a detective's desk in handcuffs. He used his one phone call to call Samantha at school again.

"Hello," she answered, annoyed.

"Samantha! Oh, thank god! You gotta help me."

"You cannot keep interrupting my classes."

"I got arrested this time!"

"Well, your plan is shot. I don't know what you want me to do."

"Have you seen Hannah? And the journal?"

"I told you I wasn't there. I did get the journal though."

Chris was stunned for a moment. "You did? How?" She didn't answer. "Samantha, you have to get me out of here!"

"I don't think that's going to be necessary."

"What do you mean?"

"All of this time travel is really messing with your memory. Give it some time. It's all going to be fine." She eyed Hannah's journal., "I've got some reading to catch up on. Bye, Mr. Chris."

"Samanth—!" Chris dropped the phone on the desk and his head into his hands. The detective came back, hung up the phone, and grabbed Chris's arm.

"Let's get you settled in. This is going to take a while."

Samantha was sitting on the steps outside school. She was reading Hannah's journal. Young Chris approached her from behind. She seemed to be expecting him, though. She read aloud from the journal. "It's Easter . . . Wow. I am lying out on the middle of the yard. The Anderson's dog Filby is smelling me, well was. I hear a lawn mower in the distance along with a little music." Young Chris sat down next to her. She closed the journal on her finger, holding her place. She gestured at the air with it. "What we have here is a dreamer. Someone completely out of touch with reality."

XIX.

Thaddeus and Luna were outside smoking, doing their usual riffing on passers-by.

"I'd like to be there for her Initial Public Offering," Luna started.

Thaddeus laughed, "Good one." Then he added, "I'd invest in her Silicon Valley!" They smoked for a moment before another walked by. "I'd start an LLC company with that one, a Limited Liability Copulation."

"I'd get caught insider trading with her!"

"You're on a roll!"

"I'm a business minor," Luna reminded him.

"Ah, yeah."

Dean Davis, dressed in a smart pantsuit, stuck her head out of the door. "Good afternoon, Mr. Noble."

Startled, Thaddeus dropped and stamped out his cigarette. "Hi, Dean Davis. How are you?"

"Have you seen Christopher Bridges?"

"No, ma'am. I believe he's sick."

"I see. Are you sure? I hear he's missed more than a few classes lately, and he didn't show up for his prospectus defense this morning."

"No, ma'am. I am not sure."

"Well, let me know if you hear anything."

"Yes, ma'am."

She disappeared back into the building.

"Chris, where are you?" Thaddeus wondered aloud.

XX.

Chris couldn't sit still. He lay down, got up, paced, lay down again. He was fuming and worried.

He looked at the clock outside his cell, looked at the ceiling. Looked at the clock, looked back at the ceiling. Kicked the bed, bounced in a fit, almost screamed, looked at the clock, then back at the ceiling.

Finally, the room started to ripple, and Chris disappeared from the cell in a bright, white flash of light.

XXI.

Chris landed naked on the floor of the closet on top of his clothes, watches, and ring from the day before. Again. All the lights in the house were off again. The generator ran out of gas, the fuse blew, and he was sucked back to the present. He'd only missed only a few hours this time, though.

He scrambled to his feet, clutching his clothes. A confused look crossed his face when he didn't see Hannah's journal. He dug around a bit, but it seemed to really be gone.

The phone rang in Thaddeus's empty office. He unlocked the door and stumbled into the office, dumped his books, student papers, notes, and keys on his desk, fumbling for the phone, "Hello?"

"Hey."

"Chris?! Dude! Where have you been?"

"You know where I've been."

"No, I don't! You've missed some really important shit lately, man!" Chris didn't say anything. "Let me guess, you went back with a plan, but the plan didn't go as planned?"

"Yeah, the journal's gone, but I have to go back."

"No way, man. Whatever you're doing, you're not disappearing on me again. Besides, if the journal's gone, as far as I can figure, your problem is solved."

"Yeah, but Samantha has it. It didn't get returned to its rightful owner!"

"She'll take care of it!" Thaddeus pleaded. "Well enough, Chris. Leave it alone."

"You know I can't."

Thaddeus shook his head. "I know."

XXII.

Samantha and Young Chris were sitting on the steps.

"You've got to give it back," Young Chris pleaded.

"You've got to give it back!" Samantha countered.

"I told you I can't."

"Why not?"

"It's been too long."

"It's only been a couple of hours!"

"I can't do it."

"We'll go together."

"Do you know her?"

"I don't know her, but I'm not scared of her." They sat in silence for a long moment. "Look, I didn't read much of this but . . ." she gestures with the journal. "There's some dark shit in here. I'm more afraid of what happens if she doesn't get it back."

XXIII.

Thaddeus and Luna were outside the physics building smoking.

"Are you going to see that blonde again?" He asked, blowing out smoke.

"Sarah? I'm not sure. Probably."

"You don't seem very excited by her now."

"Kinda passive for my taste, but I'll probably call her later."

Thaddeus stubbed out his cigarette, "Whatever bends your gender!" He pulled out another cigarette, offered her one, and turned suddenly serious. "Do you think you'll ever find one?"

"One what?" Luna asked.

"A lady. A mate. A soulmate."

"No way, man. I don't think we're built for that. I think you and I are FOMO-sexuals."

"What?"

"FOMO-sexuals. No matter who we're with, we're always going to be wondering what we're missing out on with someone else."

"That's a terrible way to frame it," Thaddeus said, taking a long pull from his smoke.

"It's the same picture no matter how you frame it!" Luna said. Thaddeus thought about it for a minute.

"Chris has been missing some important meetings the past couple of days, including his prospectus defense. I wonder what he's been doing," he said.

"Aren't you the one who would be most likely to know that at this point?"

"I guess so. Maybe we should go by his house later and see what's up."

XXIV.

The school day was over and Lubbock High was mostly deserted. Even so, Chris didn't see Principal Carter and two police officers approaching him from behind.

"C'mon, Mister. This is getting old." Principal Carter reached for Chris's arm, but he bolted, running for the Art building.

Chris rounded the corner of the school just in time to see Samantha and Young Chris talking. He saw Hannah exit the Art building and approach the bench.

"There she is!" said Young Chris.

"Okay, let's give it back."

"Give it here. I'll do it."

"Are you sure?"

"Yes."

"I don't want you running off and—"

"Give it here!"

Samantha opened her backpack and handed the journal to the frantic Young Chris.

Samantha watched from across the courtyard. Young Chris walked over to Hannah. He handed her the journal, and she hugged him in appreciation. They stood and talked for a minute. She wrote something down and gave it to him, hugged him again, and walked away.

Young Chris returned to Samantha just as blue lights began to flash against the buildings.

Chris was watching so intently from the corner of the main building that he didn't notice the lights or the police approaching from behind.

Young Chris and Samantha stood aghast as the police apprehended Chris.

Chris was lying on the bunk in the jail cell completely still. He comically looked at his wrists where his watches should be, looked at the clock outside his cell, looked at the ceiling. Looked at the clock, looked back at the ceiling.

He was giddy, worn down dumb by the events of the day.

He lay akimbo on the bed. Knowing the machine's power situation at home, he was expecting to disappear at any moment.

Finally, the room started to ripple, and Chris disappeared from the cell in a bright, white flash of light.

He was smiling.

XXV.

Thaddeus and Luna crept around to the back of the house. Thaddeus was about to check under a potted plant for the spare key, and stopped dead, noticing anew the ring of dead grass, in a perfect circle, in the middle of the otherwise fairly well-sculpted backyard. "What in the—" Luna said. They stared for a moment, before returning to the spare key.

"Same picture, different frame," Thaddeus shrugged.

Chris landed naked on the floor of the closet on top of his clothes, watches, and ring from the day before. All the lights in the house were off again. The fuse blew, and he was sucked back to the present. He'd only been gone for a little while, though. He got to his feet, picking up his clothes. A relieved look crossed his face when he doesn't see Hannah's journal. He dug around a bit, but it was still gone.

He heard the back door downstairs open and voices calling his name.

As Thaddeus and Luna entered the back door, they heard a loud *thunk* upstairs. They were startled but immediately called out for him.

"Chris!"

"Mr. Bridges?"

"Yeah! Hold on," Chris called back.

As Chris went to his bedroom to get some clothes, Thaddeus and Luna looked around downstairs. The house was completely in order. The books were all put away neatly; every surface and every item were tidy and clean. They were baffled. Chris came down the stairs, also slightly astonished by the cleanliness of the house. "Did you guys clean up in here?"

"The spare key under the fern out back—" Thaddeus answered. "Oh. No, we just got here."

"Did you see the Fairy Ring?"

"Is that what that is?"

"It's a Type 1 Fairy Ring—or Fairy Circle, if you prefer."

"I don't have a preference! Where the hell have you been?"

"They're caused by mushrooms," Chris continued, "and sometimes a weird collaboration between mushrooms, rabbits, and the grass, of course." He cut his eyes at Thaddeus.

"Did you manage to set things straight this time?"

"It feels that way."

Everyone turned toward the kitchen as keys jingled in the side door. They all looked at each other.

Chris's wife walked in carrying the mail. She flipped through several envelopes, not looking up. "What's for dinner, Mister?" She put the mail down on the kitchen counter and looked over the island at the congregation in the living room. "Oh, hey, Thad." Then she turned to Chris. "How about a pizza? I know you probably have work to do."

"Not really! But pizza sounds good."

She picked up the phone off the counter and dialed the number for the pizza place. "You guys want to stay for dinner?" They didn't answer, staring, still baffled by her presence. She gave Chris a sly smile.

XXVI.

Chris and his wife were lying in bed, having an obviously post-coital discussion. "When did you fall in love with me?" Chris asked.

"What makes you think I did?" Chris tickled her until she busted out laughing. "Okay! Okay!" She settled down again. "I think it was the second time you showed up in my house naked."

"Really?"

"Yeah. There was something about that second time, a familiarity I had never felt before." She smiled at him. "You still owe me for that hotel room... Wait... Whoa... Wicked déjà vu!"

Chris smiled back.

XXVII.

The next morning, Young Chris climbed onto the bus as it idled by the curb. Young Thaddeus had saved him a seat. Just before they got to school, Young Thaddeus turned to Young Chris.

"Look. This is going to sound weird, but—" Thaddeus started. Young Chris looked at him. "A lady came to school to see me yesterday. I'd never seen her before, but she told me to look out for you." He swallowed against his rising nerves. "She told me a lot of weird things, but mainly she told me to look out for you."

"What does that mean?"

"I don't really know." The bus stopped and parked, and students started lining up for the door. "Just . . . Just let me know if you ever need anything, okay?"

XXVIII.

The pile of mail lay on the counter in the dim morning light. A few of the envelopes were addressed to both Chris Bridges and Samantha Bridges.

One was addressed to Samantha's maiden name, Samantha Sever.

Acknowledgements

THANKS TO THOSE WHO HAVE SUPPORTED THESE and other forays into fiction, to Craig Douglas, Nicholas Mitchell, Rasta Musick, Alan Good, Josh Chaplinsky, Paul Levinson, Will Wiles, Gary J. Shipley, Charles Yu, Tim Maughan, Matthew Battles, Alyssa Byrkit, Alec Cizak, David Leo Rice, Scott Nicolay, Ira Rat, Matt Schulte, Austin Tolin, Sean Croghan, Matthew Revert, Mike Corrao, Gabriel Hart, C.W. Blackwell, J. Matthew Youngmark, Jeph Porter, Peter Relic, Claire Putney (from whom I stole the title, "Fender the Fall"), Michelle "The Little One" Andino (from whom I stole the title, "Different Waves, Different Depths"), Jaqi Furback, Cynthia Bayer, Danika Stegeman LeMay, Nicole NeSmith, Erin Cullins, Christina Zimmerman, Tina Monaco, Taylor Pecori, Rachel Carter, Shannon Keane, Eleanor Purcell, Sidney Brinson, and Claudia Dawson for their time, feedback, and kind words.

I owe a great deal of thanks to friends who let me quote and misquote them about the fictional band Bad Flag: John Mohr, Nils Bernstein, Steve Albini, Corey Rusk, and Erik Hoversten. Bad Flag was loosely based on Erik's band from the 1990s, so if you're wondering what Bad Flag might sound like, check out A Minor Forest.

I am also fortunate to have worked with Patrick Barber again. His eye for layout and design has enhanced my written work for decades now. Extra special thanks to Patrick for believing in these stories and for being so open and collaborative. Many thanks also to Mike Corrao for his typesetting job on the original "Fender the Fall" novella, and to Jeffrey Alan Love for the amazing cover art.

And as always to Lily Brewer, I cannot even make it up!

Publication Credits

“Drawn & Courted” (as “Façade”), “Kiss Destroyer” (as “The Destroyer of Worlds”), “Antecedent,” and “Not a Day Goes By” were all originally on the Close to the Bone Publications website in 2020–2021: www.close2thebone.co.uk

“*Dutch*” was in the anthology *Pontoon*, published by Malarkey Books in 2022, and published online in 2021 by *Revolution John*.

“Hayseed, Inc.” was in the anthology *Cold Christmas and the Darkest of Winters*, published by Cinnabar Moth in 2021.

“Fender the Fall” was published as a stand-alone novella by Alien Buddha Press in 2021.

About the Author

ROY CHRISTOPHER is an aging BMX and skateboarding zine kid. That's where he learned to turn events and interviews into pages with staples. He has since written about music, media, and culture for everything from books and blogs to national magazines and academic journals. He holds a PhD in Communication Studies from the University of Texas at Austin. This is his first collection of short fiction.

Printed in the USA
CPSIA information can be obtained
at www.ICGtesting.com
LVHW021218021023
759759LV00068B/1246